Consumer Guide to Solar Energy

Easy and Inexpensive Applications for Solar Energy

Scott Sklar

and

Kenneth G. Sheinkopf

Bonus Books, Inc., Chicago

95 94 93 92 91 5 4 3 2 1

International Standard Book Number: 0–929387–23–6
Library of Congress 91–71893

Bonus Books, Inc.
160 East Illinois Street
Chicago, Illinois 60611

Printed in the United States of America

Contents

Note to Readers

A portion of the proceeds from the sales of this book is being donated to Americans for Clean Energy (ACE), a national nonprofit organization designed to inform the general public of ways to utilize alternative energy systems to save energy and help the environment, and to the Solar Energy Research and Education Foundation (SEREF), a nonprofit organization which sponsors and coordinates many educational activities. For information on ACE and SEREF, write to Post Office Box 1725, Washington, D.C. 20013-1725.

Foreword

At a time when an increasing segment of the world's population is enjoying the many benefits of today's modern technology, the state of our environment is rapidly degrading around us. The increased burning of fossil fuels is wreaking havoc on our planet. Acid rain is killing our fish and forests. Unpredictable climate change is diminishing our food output. And people around the world are suffering increased cancers, emphysema, and a wide variety of allergies. Thus, the cost of much of this progress has been the serious destruction of the air we breathe, the water we drink, and the soil that grows our food.

The good news, though, is that individuals like you and your family can make a difference by helping improve our world, not by hard sacrifices, but through intelligent purchasing decisions.

During my forty years as an actor, I have travelled extensively throughout the world, and I have seen hundreds of examples of environmental changes—in forests, deserts, oceans and lakes, and most seriously of all, right in our cities. Though most people know me through my acting roles, I'm especially proud when people think of me as an environmentalist.

As one example of my deep concern for our planet's well-being, I recently built a new solar home in a beautiful area of Colorado to serve as an example of how solar technology, combined with recycled building materials—basically, used automobile tires and aluminum cans—can create a comfortable, practical residence with a minimum use of energy, in harmony with the natural order of things. The house uses such technologies as solar water heating, passive solar design principles, and photovoltaic cells for solar electricity—basic, proven solar applications fully described in this book—and reminds me every day of the significant advances our country has made in the use of solar energy.

We all know of the state of our environment today. We have read the newspaper headlines, seen the television specials, and heard hundreds of speakers from many diverse organizations who have voiced their concerns about the damage we have done to our planet. Students in the early grades of school are already familiar with terms like acid rain, the Greenhouse Effect, and energy crisis. In less than two decades, we have seen energy shortages that resulted in long lines at gas stations, skyrocketing utility bills, limited supplies of home heating oil, and even a Mideast war that brought hundreds of thousands of U.S. military men and women to the battlefield in a conflict centered around the availability of oil.

You've got to realize that the fight against our deteriorating environment cannot be led by governments. It cannot be led by multinational organizations like the World Bank or the United Nations. Rather, if changes are to be made in the way we are destroying our air and water, we as individuals will have to be the ones to make these changes.

The choices we face do not mean we have to reduce the quality of our lives, or otherwise lessen our enjoyment of the modern products and services that are so available today. Instead, we must make educated, informed decisions about the kinds of buildings, appliances, and energy we use in our daily lives. These decisions will not only help improve the quality of our individual lives, but will also save us money and help ensure that we and our children will have a reliable source of energy and a clean, safe environment in the future.

How we continue to treat our environment will affect our

lives for many years to come. According to *The State of the World, 1991,* by the prestigious Worldwatch Institute, there are a number of important environmental issues facing us today that must be dealt with. Consider that each year, as lumber and firewood are harvested and forests are cleared for farming, the planet's tree cover diminishes by some 17 million hectares—an area the size of Austria. Grazing land is being degraded in North America, Australia, and throughout the Third World, with annual losses of topsoil from cropland estimated at 24 billion tons—about the amount of Australia's wheatland. The principal greenhouse gas in our atmosphere, carbon dioxide, is increasing 0.4 percent per year from fossil fuel burning and deforestation. Air quality has deteriorated to the point that pollution has reached life-threatening levels in hundreds of cities, and crops have been damaged in countries all over the globe. And finally, rising temperatures and ozone layer depletion are adding to the habitat destruction and pollution that are reducing the planet's biological diversity. It is very difficult to read about these effects on the earth and not want to do all that we can to help the problem.

In the following pages, Scott Sklar and Ken Sheinkopf show you dozens of practical uses of solar energy—choices you can make which will actually pay for themselves while reducing the pollutants in our environment and helping cut down on further damage to the planet's eco-system. There are dozens of steps you can take to cut down on your energy use while helping keep the air and water clean—without losing any living quality and actually helping you to improve your personal comfort and your standard of living. Indeed, these choices are more than just a simplified set of options. They are instead, a set of low-cost applications that, when undertaken by a large segment of the general public, will have positive and profound environmental consequences for all of us.

In places from Japan to Israel, Australia to Greece, solar water heating systems are used everywhere. If we could just increase our use of solar energy in this country to the level used by the Japanese, for example, we could save nearly 4 percent of the total U.S. energy consumed in the next decade. Increased use of solar electric cells could provide enough power to supplement another 6 percent of our country's energy. I use both of these technologies in my home, and I know how efficiently they work, how reliably they perform, and how cost-effectively they do their job without harming the environment.

Residential and commercial buildings in the U.S. currently use more than 35 percent of our energy. The solar technologies explained in this book can help consumers make a significant impact on their energy use. All told, the 1.2 million solar water heating systems in our country are saving 9 million barrels of oil every single year.

In this book, the authors lay out a coherent and practical pathway for the use of solar technologies. If each of us does what we can and follows the pathway, our nation and our world will be a far better place. The *Consumer Guide to Solar Energy* shows you the steps you can take to make your choices.

Human beings are an interesting breed. We copy each other, so if you want to make a healing impact on the one home that we all share, start taking the actions that remove the stress and trauma that we have heaped upon this earth. Walk your talk and you will be amazed at the number of people that will follow in the tracks that you make.

Yours for a cooperative, just, and peaceful world.

Dennis Weaver

Dennis Weaver
Actor and Environmentalist

Preface

August 2, 1990. The world woke up to a Mideast conflict and the start of another energy crisis. For the first time since the 1970s, we started worrying again about the rising price of gasoline, the availability of fuels to heat our homes as well as to power our cars, and the possibility of fighting in the Middle East. Daily news reports showed oil prices climbing from below fifteen dollars a barrel to more than thirty dollars a barrel, with forecasts of skyrocketing prices on the horizon.

Iraq's invasion of Kuwait once again reminded Americans how dependent we are on oil imported from countries that are not always friendly to us. More than 50 percent of our energy is imported, and over one-fourth of the oil and petroleum products imported into our country come from the Middle East. Americans pay trillions of dollars for our imported oil supplies.

Those of us who lived through the two major oil shortages of the 1970s will never forget the inconveniences and the dangers. The lines at the gas pumps may have been annoying, but the danger of energy shortages to our national economy, the lack of heating fuels in the coldest days of winter (with potential health hazards for the children and senior citizens in our homes), and the cold showers and baths, electric "brownouts," and other power interruptions caused serious problems for millions of Americans.

So what did we learn from the shortages of the 1970s? Apparently very little. By the end of the 1980s, Americans seemed to have gone back to their ways of old. Huge gas-guzzling automobiles were returning to our highways as car manufacturers brought out bigger models. Residential energy use started climbing upward again as people ignored appliance "EnergyGuide" labels and instead bought appliances without regard for their efficient energy usage. Realtors were told by home buyers that they cared more about the color of the carpets and the finish on the kitchen cabinets than they did about the energy efficiency of the home. Sales of solar equipment and other types of renewable energy systems plunged as government tax credits and other incentives were allowed to expire. And maybe most serious of all, our country turned more and more to imported oil to meet these growing power needs.

For most of its history, the United States had produced its own fuels domestically, and we had plenty of petroleum, coal, wood, and natural gas. By the early 1970s, though, the fuel of choice—petroleum—had reached more than 50 percent of the mix. Just a 3 percent cut in these oil imports during the embargo wreaked havoc in our society by causing long lines at the gasoline pumps and immense increases in our monthly utility bills. Even after all this, we still turned to petroleum products for more than 40 percent of the energy used in the U.S. last year.

The effects of the energy crises of the 1970s started changes that have brought massive shifts in the world's balance of power. In less than twenty years since that time, the greatest transfer of wealth the world has ever seen has occurred. Even when the United States was the richest nation in the world, we were sending our hard-earned dollars overseas to buy petroleum, which we burned up.

Today, the United States is the largest debtor nation in the world. We have gone from being the world's richest country to the third-richest (behind Germany and Japan). Oil imports still account for the largest component in the U.S. trade deficit. Unless our nation begins to use domestic energy resources, we will become poorer and poorer.

The cleanest, easiest, and most reliable fuel source we have is solar energy. Yet, Japan, Germany, Israel, Australia, and other countries rely on solar energy more than we do. In fact, even as we boast of the fact that one million homes in the U.S. have solar energy systems, the city of Tokyo alone has more than 1.5 million solar buildings.

But there is another serious problem. It appears that even as we are losing our wealth, we are losing our planet. The carbon from burning natural gas, petroleum, and coal is widely acknowledged as a major cause of global climate change. If carbon is continually pumped into the atmosphere, the resulting change in our climate will ruin food crops, forests, and our shorelines. The sulfur from burning coal and petroleum causes "acid rain," which kills fish, forests, and causes major health problems ranging from allergies to cancers.

But pollutants and the "greenhouse effect" can be mitigated if people act now. Solar energy is a simple, efficient way to maintain America's economic strength and ensure the world's environmental health. Every American can make a modest investment in solar energy for water heating, building heating, lighting, and ventilation and actually realize economic gain and improved home comfort, without changing his or her lifestyle. We don't need to sacrifice in order to contribute to our world. But we do need to make some critical decisions now, with a close look at the long term and the world we are creating for our children and grandchildren.

Sure, it may be more convenient for you to call the local plumber and replace your conventional water heater when it breaks, or ask the electrician to run wiring through your back yard for lighting your walkways or patio. But you owe it to yourself, as well as to your society, to consider your solar options. This book is intended for those with a will to act on the personal level to save money for themselves while they are helping provide reliable energy now and for the future. At the same time, they are taking a small step in turning the U.S. economy around and helping to preserve the health and life of our planet.

The following pages explain how the various types of solar energy systems work, how to buy the right equipment for your needs, what questions to ask and information to collect, and how you can use this free energy source to keep your home comfortable and energy-efficient.

In many ways, because the technology is mature and ready to use today, the questions are more difficult than the answers. It is hoped that this book will stimulate you to ask how you and your family can take advantage of solar energy to heat the water in your homes, provide power to lights, fans, and other equipment, extend your pool's swimming season a few more months each year, enjoy a more comfortable home with lower monthly utility bills, and otherwise let renewable energy from the sun help cut down on your use of fossil fuels. Most of the electricity we use today is generated from oil, petroleum, or nuclear power—all of which have finite resources and severe waste and emissions problems.

As environmentalists wrestle with the problems of acid rain and the greenhouse effect, as our nation's leaders wonder how to decrease our huge dependence on foreign oil, as another Mideast war threatens our oil supplies, and as we ponder the prospects of recurring energy crises with rising prices and certain shortages, we ought to realize that the answer to these problems has been right above us all this time.

Solar energy is a bright idea whose time has come. Now we will show you how to put it to work for you and your family.

Acknowledgments

The authors gratefully acknowledge the support and assistance of the experts in the solar energy industry and research community in the United States during the research and writing of this book.

Many members of the industry provided helpful information, made comments on drafts and ideas, and otherwise provided a great deal of technical and general information on the solar technologies. We salute the solar industry, and the members and staff of the Solar Energy Industries Association (SEIA), who are helping make solar energy a reality.

Special thanks must go to Dr. David Block of the Florida Solar Energy Center and the members of his staff who provided a great deal of background information and assistance. Much of the material in this book has been drawn from their research and technology transfer efforts. Thanks are also due the staff at the Solar Energy Research Institute in Golden, Colorado, and Sandia National Laboratories in Albuquerque, New Mexico, for their ongoing efforts to develop and commercialize solar energy systems.

In 1991, the solar industry in the United States celebrates its 100th birthday. During the past century, tremendous technological advances in materials, science, and knowledge have made it possible for consumers to use the sun's power for many applications in their homes, automobiles, and boats, and for other uses. We are tremendously indebted to the members of the industry, past and present, whose efforts have resulted in proven, reliable products that are available today for hundreds of uses. The work of dozens of these people has been useful in preparing this book.

Finally, we acknowledge the advice, understanding, and patience of our families during the many months it took to compile this information. Scott Sklar's wife, Kathy Zwicker (the computer widow), and Ken Sheinkopf's wife, Blanche, and sons Adam and Jeff tolerated our many late night and weekend retreats to our offices for writing, editing, researching, and more writing and editing.

The result is a guide you can use to help start saving money, energy, and our environment. It's not really a big job, but you've got to take that first step. We hope this book gives you the push you need to make that step.

About the Authors

Scott Sklar is Executive Director of the Solar Energy Industries Association and the U.S. Export Council for Renewable Energy, a consortium of the eight energy efficiency and renewable energy trade associations promoting exports. He formerly served as Political Director of the Solar Lobby and Washington Director and Acting Research Director of the National Center for Appropriate Technology. He began his career in energy by serving as an aide to Senator Jacob Javits (NY) for nine years. Mr. Sklar is the author of numerous books, articles, and Congressional testimonies on solar and renewable energy. He currently is a member of the U.S. Department of Energy Advisory Committee on Renewable Energy and Energy Efficiency Joint Ventures. He lives in a home in Arlington, Virginia, with solar water heating, solar space heat for a greenhouse, and uses photovoltaics to charge his electric car.

Ken Sheinkopf is Director of Special Projects for the Solar Energy Industries Association and Director of Governmental Relations for the Florida Solar Energy Center. He also serves as Executive Director of the American Ocean Energy Industries Association. He holds B.A. and M.A. degrees from Syracuse University, and did advanced graduate work at the University of Wisconsin. Mr. Sheinkopf has been a faculty member at the Pennsylvania State University, Rollins College, and the University of Central Florida, and has been active in the solar energy field since 1983. He has given more than 200 presentations on solar and renewable energy at meetings throughout the world, and written dozens of articles for energy journals. Since 1986, he has written a weekly column on energy efficiency and home comfort for more than twenty newspapers and magazines. He currently lives in Potomac, Maryland.

CHAPTER ONE

Introduction

Solar energy is ready for use today

It has been estimated that 1,000 times more energy reaches the earth's surface from the sun every single year than could be produced by burning all the fossil fuels mined and extracted during that same year. Today's solar energy equipment offers a cost-effective, efficient, and reliable way to take advantage of that potential power.

In 1891, an inventor in Baltimore, Maryland, patented the first solar water heating system manufactured in the United States. Since that time, a major industry has grown in the United States—slowly at first, in California and Florida, then more gradually into the Midwest and New England, and finally into the rest of the United States. There have been

Solar systems have been used in the U.S. for the past 100 years, and have become very common in American homes (*American Energy Technologies, Inc.*).

booms and busts during the past century—times of great public interest, government support, and product innovation. And there have been times of declining interest, technical problems to solve, and lack of consumer awareness.

Today, as the solar water heating industry completes its first century of operation, the prospects of utilizing this technology are as bright as the sunshine on a hot, summer day. Hundreds of solar manufacturers, distributors, component suppliers, contractors, and researchers, from coast to coast, are working in an industry whose time has come.

The products have been tested, certified, and widely used. The technology has matured to the point where more than one million Americans use solar water heating systems to provide for most of their hot water needs each day of the year. Hundreds of thousands of commercial and industrial buildings—from car washes and laundries to office complexes and hotels—use solar water heating systems to reduce their electric, fuel oil, or gas usage and provide the hot water their customers need.

Another solar technology that is becoming more popular and economical produces electricity rather than heat. In the early 1950s, United States' scientists were confronted with

Supplying utility power to locations like this bus shelter can cost thousands of dollars for the cable and conduit. Instead, a photovoltaics system providing light to the shelter can be easily installed and relocated as needed at very low cost (*Photovoltaic Energy Systems, Inc.*).

PV panels are used for many commercial and industrial tasks. This microwave repeater station uses solar power instead of relying on a diesel generator which would require bringing fuel to this remote location (*Photovoltaic Energy Systems, Inc.*).

the problem of finding a power source for the space satellites being developed. They turned to photovoltaics—solar cells made from silicon, a very abundant element on our planet. In time, civilian companies developed commercial uses for these cells including powering consumer products such as calculators, watches, and cameras.

In just the last five years, though, research and manufacturing breakthroughs have reduced the costs of solar cells while improving their efficiency, and their uses have multiplied. For the homeowner this means that instead of paying the high costs of running electric wires to the attic or the back porch, photovoltaic panels can be used to power vent fans and security and walkway lights, charge batteries in vehicles, and provide power for many home appliances.

Not too long ago, the solar cell was a novelty. Today it is a serious, useful product that can meet many consumer needs. Prices of cells have dropped more than a hundredfold since the 1970s to the point where solar electricity is practical and cost-effective in many uses today.

Other uses of solar energy abound. Solar systems provide economical heat for swimming pools, spas, and hot

Students from thirty-two U.S. colleges and universities designed and built solar-powered cars for the 1,800-mile Sunrayce held in 1990. While cars like this one won't be widely used on the roads in the near future, they do show the reliability and versatility of this power source (*Adam Sheinkopf*).

Solar systems fit in well with any type of architecture (*Solar Development Inc.*).

Many educational efforts keep consumers informed of solar products and services (*Florida Solar Energy Industries Assn.*).

tubs. Consumers who spend leisure time away from home are using solar power to purify drinking water, cook their food, and keep refrigerated food and drinks cold. Builders, architects, and designers are using passive solar principles to make today's houses more comfortable and more energy-efficient.

As a result of the development of these technologies, the question facing consumers in America today is not "when will solar energy be available to us," but, rather, "how do we take advantage of it today?"

Significant technological advances by industry and research labs during the past decade have brought the equipment to a new stage of reliability and performance. Educational efforts by government and industry groups have trained thousands of installers, service people, home inspectors, appraisers, and others involved in the installation, use, and purchase of solar energy equipment. Consumers are finding that solar chargers mounted on the dashboards in their automobiles are keeping their batteries in peak working condition, while outdoor lights are providing safety and security to areas too expensive to power by running electric wires over even short distances.

Solar energy is a safe, reliable power source that you can use today.

CHAPTER TWO

History

A look at 2,000 years of solar energy

Ask people what they think of solar energy, and many of them will tell you that it's probably going to be important in the future. After all, many people feel solar energy is something pretty new that grew out of the oil embargoes of the 1970s. While it is true that there has been significant progress in the development of solar equipment and energy-saving strategies during the past twenty years, the use of the sun to provide power for our needs goes back more than 2,000 years.

As long ago as A.D. 100, Pliny the Younger, whose books on the daily life of his times are a great source of historical information, built a summer home in Northern Italy that included a windowed room using thin sheets of mica or selenite. The room got hotter than the others and conserved severely dwindling supplies of wood. In fact, it was during this time that glass was first perceived as a way to keep in heat and keep out cold, rather than a building feature to be used only for decoration. Later, from the first to the fourth centuries A.D., the famous Roman bath houses were built with large windows facing south to let in sunlight and utilize the sun's warmth.

By the sixth century, sunrooms on houses and public buildings were so common that the Justinian Code stated that sun rights were to be adhered to from then on: "If any object is so placed as to take away the sunshine from a heliocaminus [sunroom], it must be affirmed that this object creates a shadow in a place where the sunshine is an absolute necessity. Thus it is a violation of the heliocaminus right to sun." Stimulated by the fable of Archimedes using solar mirrors to burn enemy ships attempting to destroy Syracuse, many famous scientists throughout the centuries tried to devise destructive solar mirrors for military purposes.

In the 1700s, it had become common practice for the aristocracy to use fruit walls to store heat, and then release this heat to the trees planted next to them. Using this passive solar strategy, fruits would ripen earlier and the fruit-bearing time would be increased. England and Holland led Europe in the development of actual greenhouses with sloping glass walls facing south. Conservatories became very popular among the wealthier classes in the 1800s, creating places where guests could stroll through large, comfortable solar-heated greenhouses with lush foliage.

The Swiss scientist Horace de Saussure is credited with inventing the world's first solar collector or solar hot box in 1767 (see chapter 7). Renowned astronomer Sir John Herschel used solar hot boxes to cook food during his expedition to South Africa in the 1830s.

A French scientist, Augustin Mouchot, patented his solar engine in 1861. By the next decade, he had demonstrated his solar cooker, using a conical reflector, in the French colony in

Algiers. At that time, Algeria was importing 85 percent of its fuel, and the French government was looking for ways to promote solar energy. In addition to his work with solar cookers, Mouchot demonstrated solar water pumps for irrigation and even solar stills to make wine. In the late 1800s, in fact, solar distillation of water was the most widespread use of solar energy.

The post-Civil War era saw the development of solar energy in this country. The great westward movement of the pioneers after the war found people leaving water in black pans during the day so that the sun would give them heated water by the evening meal.

The history of the solar energy industry in the United States can be traced back to the work of an engineer in the early 1880s, John Ericsson, who generally is credited as being the first American solar scientist. At the time he began his pioneering solar work, he was already well-known for designing the Union's ironclad ship, the *Monitor,* which defeated the Confederacy's ironclad ship, the *Merrimac,* during the Civil War. Ericsson was obsessed with developing solar energy to power steam generators for ships. Although he did develop several solar-driven engines, he was unable to commercialize this technology before he died.

In 1892, inventor Aubrey Eneas founded the Solar Motor Company of Boston to develop solar-powered motors to displace the conventional steam engines running on wood or coal. By the 1900s he had demonstrated a large solar-driven engine in an ostrich farm in California, and opened a California office to sell engines. The first engine he sold failed, however, and his company folded.

While all of these inventors were involved in utilizing solar power, the man considered to be the real father of solar energy in the United States is Baltimore inventor Clarence Kemp, who, in 1891, patented the first commercial solar water heater. In 1895, two businessmen bought the rights to his invention and established a company in Pasadena, California. Since gas and other fuels were fairly expensive at the time, the product quickly became very popular and nearly 30 percent of the houses in Pasadena had solar water heating systems by 1897!

The people marketing Kemp's Climax Solar Water Heaters knew that they had a captive market. Both coal and synthetic gas had to be imported to California, and, in terms of 1890 dollars, were more than six times more expensive than energy is today. According to historians Ken Butti and John Perlin, the Climax Solar Heaters sold for $25 (which is more than $300 in today's dollars). Interestingly enough, that $25 investment could be paid back in less than three years since the average family saved about $9 annually on coal costs.

A major technological advancement took place when William J. Bailey of the Carnegie Steel Company came to Los Angeles in 1908 and invented the kind of solar collectors that are the predecessors of the ones popularly used today. Bailey built a heavily insulated box with copper coils, which brought the heated water into a tank on the roof where it was gravity-fed into the house. By the end of World War I, more than 4,000 of these solar systems had been sold. In 1923, Bailey sold the patent rights to a Florida businessman, who sold nearly 60,000 of the systems by 1941. However, the infant solar water heating industry went into sharp decline during World War II when copper, a key component of the process, was heavily rationed.

Also due to the prolonged war, energy was becoming scarce and more expensive. In the United States, many Americans were looking to build more energy-efficient homes that took advantage of building techniques to hold solar heat in the home in winter and allow inside heat to escape outdoors in the summer. Using good building design, orientation, and materials in this way is known as passive solar construction. By 1947, demand had become so great for solar homes that the Libbey-Owens-Ford Glass Company published a book entitled *Your Solar House,* which profiled house plans of forty-nine of the nation's greatest solar architects. The book featured designs by architects from each of the forty-eight states and the District of Columbia who were asked to submit drawings of homes which should cost no more than $15,000 and be built to pre-war standards. This book and other publications helped spur the use of passive principles in new homes, and by the late 1940s, a large number of housing developments had been built throughout the United States using both active and passive solar applications.

While oil prices began to decline in the 1950s, reducing the upward spiral of construction starts of solar houses that had continued until the late 1940s, energy costs remained high in the growing Southwest. In the mid-1950s, architect Frank Bridgers designed the world's first commercial office building using solar water heating and passive design. The solar system has been continuously in operation since that time. In 1989, in fact, the Bridgers-Paxton Building was placed in the National Historic Register and officially recognized as the world's first solar-heated office building.

Photovoltaic cells were originally developed to provide power for U.S. space satellites and manned flights like the Skylab (*NASA*).

Coincidentally, around the same time Bridgers was designing his building (1954), Bell Telephone researchers accidently discovered the sensitivity of a silicon cell to sunlight. This was the birth of photovoltaics (the solar cell). Bell intended to use the solar cells to power remote telephones, but it turned out that the U.S. space program became the real launching pad for photovoltaics. By the late 1950s, solar cells were being used to power all U.S. satellites; they continue, today, as the prime power source for both unmanned space satellites and manned projects aboard the Space Shuttle. The large solar sails are an outstretched feature on almost all space vehicles, other than ones scheduled to go into very deep orbit where there is limited sunlight.

By the 1960s, a handful of solar companies in the United States were making solar water heaters or solar cells. But world events changed the business situation radically when President Richard Nixon allowed the administrative limit of oil imports to surpass 50 percent. By 1973, U.S. oil imports had reached nearly 60 percent and the oil-producing nations formed a petroleum producers cartel named OPEC (Organization of Petroleum Exporting Countries). The embargo that year by OPEC members held production down by only 3 percent, but that reduction in oil availability caused havoc within the U.S. economy. Long gas lines and sharp increases in petroleum prices were so severe that prices at the pump increased nearly fourfold.

By 1974, two new solar water heating manufacturers came of age: FAFCO, a California company which specialized in solar pool heating, and Solaron, a company in Colorado which specialized in solar space and water heating. This was a historic time for the solar industry. These two companies were the first national solar manufacturers in the country, and represented the beginning of today's commercial solar industry.

A couple of years later, as a result of the 1973 oil embargo, President Jimmy Carter stated that he wanted to do what he could to develop the potential of solar energy to help reduce our dependence on foreign oil. He installed solar panels on

Simple solar systems can provide economical hot water for residential use (*FAFCO*).

the White House while he promoted a wide range of incentives for solar energy systems to stimulate their use.

By the time the second oil embargo hit the United States in 1979, the Solar Energy Industries Association (SEIA), the national solar manufacturers' association founded five years earlier, had grown to more than 30 member companies. Within the next couple of years, the U.S. solar industry had grown to more than 100 national solar manufacturers and component suppliers producing solar water heating, solar thermal electric, and photovoltaic equipment. Now, more than 400 other companies are members of SEIA's four state chapters. Included in this roster are many contractors, installers, and solar maintenance companies.

During the 1970s, a broad range of federal and state incentives were established to promote solar energy. A few of these are still around today, even after the lower energy prices of the 1980s had winnowed away many of the energy incentives that were originally created to meet the threat of increased oil imports.

On Nov. 9, 1978, the Public Utility Regulatory Policies Act (PURPA) was signed into law. The law is one of the most significant pieces of legislation in the development of the solar industry. PURPA marked the first time that an individual or private company had the legal right to hook-up to the utility company and also be paid a fair rate for the excess electric power generated by the solar system, which was sold back to the utility. PURPA is still in effect today, and the state utility regulatory commissions are required to establish the rules of interconnection and how to best determine the fees to be paid. The federal law requires that the principle of "avoided cost" be used to determine rates. These costs are the expenses of building a new electric power plant which would be avoided if the utility company used solar energy instead of constructing a conventional energy plant. PURPA has been amazingly successful. There are now more than 10,000 homes in the United States powered solely by solar energy. In California, there are solar steam-to-electricity plants producing more than 400 megawatts of power—enough to provide for the energy needs in more than 400,000 homes.

Another of the critical pieces of legislation affecting the development of the solar industry was started in 1978, when solar businesses and individuals were allowed to take a credit on their federal income tax on a percentage of the cost of the installation and purchase price of a solar water heating, solar space heating, or solar electric system. By 1982, a 40 percent residential solar tax credit and a 10 percent business energy investment credit were both in effect. At the end of 1985, though, the residential credit was allowed to expire as part of President Ronald Reagan's Tax Reform Act. However, the solar business energy tax credit has been extended every year, and is currently set to continue through Dec. 31, 1991, with industry hopes for extension beyond that date.

How successful were the tax credits in stimulating the growth of the solar industry? Consider that in 1978, 30 national solar manufacturers with less than $30 million in annual sales existed. By 1985, nearly 150 solar water heating manufacturers with more than $800 million in annual sales were in operation. Today more than 1.2 million buildings in the United States have solar water heating systems. There are also 250,000 solar-heated swimming pools, 10,000 solar electric homes, and nearly 300,000 buildings with major passive solar features.

In addition to the federal incentives, thirty-seven states also had state tax credits which in most part were tied to the federal residential credit. Today there are still twenty-four states with some type of incentive for solar purchasers, but most of them consist of lesser incentives such as sales tax exemptions or state property tax waivers so that the homeowner or business does not have an increase in property tax solely because of the installation of an energy-saving device.

From 1970 to 1980, significant developments occurred in the industry. The United States Department of Energy was formed to administer solar research and development programs. Such national laboratories as the Solar Energy Research Institute and Sandia National Laboratories became active in solar energy research and development. A Solar Energy and Energy Conservation Bank, administered by the U.S. Department of Housing and Urban Development, was created to provide low-interest loans to homeowners. Unfortunately, though the Bank was signed into law as part of the Energy Security Act of 1980, President Reagan's energy budget cuts virtually shut down the program. During this same period, a number of states created alternative energy research, development, and education centers. This included the Florida Solar Energy Center, which is the largest state solar center today.

In 1983, the Wisconsin Supreme Court upheld the first solar access law based on Old English Law of the "right to light" for urban gardens. Arizona and Michigan soon followed the example. Many localities have since enacted solar access protection laws as a result of the vanguard approach by these three states.

One interesting situation in the 1980s represented a meeting of the old and new in solar. For many years, the federal government had spent millions of dollars to provide all-electric housing for the Navaho and Hopi Indians in Arizona and New Mexico. High energy costs of air conditioning during the midday and space heating at night, though, forced many of the native Americans to flee these newer homes for their more comfortable ancestral homes. In fact, the cliff homes built of clay hundreds of years ago enjoy the optimum passive solar properties of holding heat within the building materials during cool weather and maintaining cool temperatures inside during hot weather. Such passive solar properties use no moving parts and require no fuels.

Today, thanks to government help and private industry involvement, the largest residential use of solar electricity in the world is on these same Navaho and Hopi reservations, where several thousand people live in buildings powered by photovoltaics.

While the United States is a leader in solar energy, other nations during the past decade have also made significant strides in the use of the technology. In Japan, for example, there are nearly 1.5 million buildings with solar water heating in the city of Tokyo—more than in the entire United States. In Israel, about 30 percent of the buildings use solar

In Africa, this school and community center in Gabon uses PV power for radio, TV, and VCRs to provide educational materials. PV-powered lights also make possible evening classes for adults (*Photovoltaic Energy Systems, Inc.*).

water heating, and all new homes are required to have solar water heating systems. Greece, Australia, and several other countries are also ahead of us in their use of solar energy.

Many great nations in the past have believed that their conventional fuel supplies would last forever. For hundreds of years, wood was the primary fuel in Europe and North Africa. Today, forests have been cleared to such an extent that, especially in places like North Africa, we have created one of the world's largest deserts.

By the 1800s the world moved toward coal and built vast railroads to carry this fuel supply from its source to its use. Whale oil and turpentine were used for the primary source of lighting until the discovery of petroleum in 1859 added a new fuel supply. But it is clear that oil is a finite resource which will increase in price as it becomes harder to find. Experts at the American Gas Association, for example, tell us that our supply of natural gas will last only another sixty years. The burning of coal will become even more severely limited due to its acid-rain-causing emissions as well as its potential to change world climate. Nuclear energy is a very expensive source of power, and we still have not dealt adequately with the nuclear waste disposal issues.

Using solar energy gives us independence from America's reliance on imported oil (*American Energy Technologies, Inc.*).

So just as wood, whale oil, turpentine, and other early fuel sources have gone away, so will the finite resources such as natural gas, coal, and petroleum become more dear in cost and availability. Solar energy and other renewable energy sources (especially wind, biomass, geothermal, and hydropower) will begin to play a far more prominent role in the U.S. energy mix. The United States already relies more on these renewables (11 percent) than it does on nuclear power for its energy needs. It is clear that we will—and must—use them even more in the future.

On the front of the National Archives building in Washington, D.C., is a quote that reads: "What is past is prologue." This appears to be an accurate way to describe the solar energy industry, since, after 2,000 years of development, solar energy's time has arrived. It is now here to stay.

CHAPTER THREE

Solar Water Heating

A modern solution to an old problem

There is a solar contractor in northern Colorado who gives out keychains at local home shows, community meetings, and other gatherings. "You're wasting thousands of dollars every single year," the message on the keychain reads. "Why not put the sun to work for you and use all that free energy?"

His message is a simple one. Just about every day of the year, in just about every part of the world, the sun shines—sometimes very brightly, other times partially blocked by clouds. But most days, the sun is visible. If you think about all that sunshine falling on your house during the year, you realize there is a lot of energy that you ought to be putting to work.

You can also imagine the power of the sun by thinking about what it is like sitting in a hot car on a sunny day. The sun's rays enter through the windows and heat up the car's interior. That's solar energy at work, but in a very inefficient type of collector.

Solar water heating is a proven, reliable technology (*FAFCO*).

Solar water heating is the simplest way to utilize the sun's rays to save energy and money. Actually, a solar system works much like the electric or gas system you probably have in your home right now. With a conventional water heater, cold water from the main water pipe enters the tank. When a sensor in the tank detects that the water temperature is too low, a heating element turns on to warm the water to the desired temperature.

A solar water heater actually has two temperature sensors—one in the tank, as in the electric heating system, and another one in the collectors. The sensor on the collectors reacts when the water temperature is low, and turns on a pump that circulates water through the pipes in the collector. The sun's rays then provide the primary source to heat the water.

The principles of solar water heating are very basic and easy to understand. Think about what happens when you leave a garden hose out in the backyard on a hot summer day. When you turn on the faucet, the water that comes out can be pretty hot. The hose is actually acting like a solar collector, absorbing the heat of the sun and then transferring that heat to the water flowing through it. That's how solar water heating systems work. Collector tubes inside an insulated box absorb the sun's heat and transfer that heat to water or another liquid flowing through the tubes. When you need hot water inside the house, the system draws this heated water in for your use.

The key point is this: you're using solar energy instead of gas, electricity, kerosene, oil, or any other type of fuel to heat the water in your home for cooking, bathing, and other uses. That energy isn't going to run out, go up in price, or spew pollutants into the atmosphere. And best of all from an economic point of view, the savings on your utility bill will pay for that system in just a few years.

Last year in the United States, about $13 billion was spent by consumers on energy for home water heating. The amount of energy used for this task might best be understood by a simple analogy. A medium-sized automobile driven 12,000 miles during the year, at an average efficiency of 22 miles per gallon, will use about 11.1 equivalent barrels of oil. Meanwhile, an electric water heater supplying 80 gallons of hot water per day (for an average family of four) will use 11.4 equivalent barrels of oil every year. You actually use more fuel to heat the water in your house than to drive your car. In fact, an electric water heater is the single biggest energy user of all the appliances in your home.

Studies have shown that water heating accounts for approximately one-fourth of the total energy used in a typical single-family home. Think about your most recent power bill. About 25 percent of that bill went for heating water. The figure is even a little higher if you have more than four people in your family (and certainly higher if you have a teen-ager who likes to take twenty minute showers!). The amount of money spent on water heating becomes even more significant when you realize that this percentage is only a little lower than the portion typically spent for air conditioning in hot climates, highlighting the fact that water heating is a year-round requirement that costs a great deal of money to provide.

Surveys of families with solar systems on their homes have found that most people are extremely happy with the performance of their system. A recent survey in Florida, for example, revealed that 95 percent of the solar system owners interviewed were satisfied with their systems, and 84 percent said they would choose a solar water heater for their next home. Only 7 percent of the homeowners had ever encountered a problem with the operation of their system, and about 90 percent of them said the problem was corrected. Systems used in Florida are typical of the systems used throughout the rest of the United States.

Because of the low purchase price and ease of installation, home builders often choose electric resistance water heating for their new homes. Natural gas systems are also popular in areas where they are available. But when you work with a builder and plan your next home, it can make a big difference to your pocketbook if the lifetime energy costs are considered as well as the initial purchase price. Researchers at the Florida Solar Energy Center have studied the potential savings of the major types of water heating systems, and found that solar energy systems offer the biggest potential savings to homeowners (table 3-1). Their studies show that solar system owners can save up to 85 percent on their utility bills over the costs of electric water heating, as opposed to savings from heat recovery units (20 to 50 percent), heat pumps (40 to 50 percent), and natural gas (59 to 65 percent).

In addition to the financial savings, there are other important reasons why people buy solar water heating systems. You are actually giving your family a form of energy insurance by making sure you'll have enough hot water regardless of future energy shortages and rising electric costs. People also buy these systems because solar power is clean, environmentally harmless, readily available, and a way to reduce the

Table 3-1. Annual energy savings comparison: alternative water heating options and electric resistance systems (*Florida Solar Energy Center*).

Type of Water Heating	Retail System Price	40 Gal per Day Use	70 Gal per Day Use	Percent Savings
Electric Resistance	$150–$350	—	—	—
Heat Recovery Unit	$600–$1000	$32–$95	$57–$166	20–50%
Heat pump	$900–$1100	$65–$95	$113–$166	40–50%
Solar	$1500–$2500	$81–$162	$142–$282	50–85%
Natural Gas	$350–$450	$97–$125	$168–$217	59–65%

outflow of U.S. dollars for foreign oil. Recent studies by the National Remodelers' Association show that the cost of a solar system is fully added to the value of a home, making the resale value of your house that much higher.

Finally, ask yourself, what other purchases have you made recently that pay for themselves in a few years and continue paying you back for many years to come? The answer is an easy one. There really aren't any.

How much money will I save with a solar system?

Solar water heating can be more economical over the lifetime of the system than is heating water with electricity, fuel oil, or propane gas. In the Sun Belt states, solar water heating systems can easily provide faster paybacks—when the savings from the system exceeds the purchase cost—and rates of return on investments of 10 or 11 percent, or even more. In just about every part of the country, solar systems will pay for themselves during the system's lifetime. In general, you can estimate that a system will typically pay for itself in as little as three or four years to around seven or eight years, depending on the geographic location, amount of hot water used, utility rates, and other factors.

Savings are even greater in new construction, so if you're having a home built or renovated, you should talk with your builder about having a solar system installed during construction. This will cut out the costs of replacing conventional water heating equipment and its extensive installation, saving you money on both the installation costs and some of the equipment costs "up-front."

In other words, a homeowner wanting to replace his current electric water heating system has already paid about $400 for the electric tank. If you put a system in your new house, you won't need to pay for that extra tank. When the solar system is included in the financing for a new home, the typical cost ranges from only $13 to $20 per month. Because the system is included in the home's mortgage, you can take advantage of the federal income tax deduction, which is typically worth between $3 and $5 per month for the solar system. Add in the fuel savings of $25 to $45 per month, and the homeowner profits from his or her investment from the very start.

The cost of an installed solar system varies widely today, depending on the materials used by the various manufacturers and the services offered by the sellers. Current average prices in the U.S. for an installed, active system that would meet the needs of a single family home with four occupants range from about $2,200 to $4,000. Compared with the cost of buying a conventional electric, fuel oil, or gas water heater, this is a much higher initial investment. But to make an accurate comparison of the true costs of the system, the cost of the fuel for a conventional heater over its lifetime must also be considered.

Using what is called a simple payback formula, you can determine the time required for a solar system to return the investment through energy saved. First, you need to determine the net cost of the solar system. Be sure to deduct any rebates or state tax incentives. In many cities around the country, utility companies offer their customers rebates for installing solar systems (it helps reduce their "peak loads" and can eliminate or postpone the construction of new power plants to meet increased demand). After figuring the net cost, calculate the annual fuel savings, and divide the net investment by this number.

Here is an example that will give you an idea of the potential savings from a solar water heating system. Assume that a family wants to add a solar water heating system to their existing home. The system they selected costs $2,750. Let's assume that the state or locality exempts their sales and property taxes. Based on current electricity costs, a family of four could save an average of $45 per month, or $540 per year on water heating costs. Their savings will pay back the cost of this system in about five years. When you consider that the average system lasts about twenty years, the savings over the remaining fifteen years will be cash in the pocket.

In the event that your utility offers a rebate or your state offers a tax incentive, this system would pay for itself in energy savings in less than the five years.

It is important to note that if the family in this example was having this same system installed in a new home they were having built, the purchase price would be reduced even more, so that the monthly energy savings would be much greater than the $15 or less they would be paying on their mortgage to cover the system cost.

Keep in mind that this example is based on today's energy rates. Any future rate increases, which by conservative estimates will rise by about 3 percent per year nationwide, will make these savings even larger.

What kind of system should I buy?

There are basically two types of solar water heaters available today. One is called an "active" system—it uses a pump and other controls to force water through the collector and back into your home. This type of system is very popular in all parts of the country, and is especially recommended for colder climates where some type of freeze protection is necessary for the system. The other major type is called a "passive" system because it relies on gravity rather than a pump to move the water. Two different types of passive systems are widely used in the Southwest and Southeast.

Active Solar Systems

You might hear people talk about "open loop" or "closed loop" systems. These terms refer to active systems where the water is heated directly for use (open loop), or where an antifreeze liquid is heated before transferring its heat to the water by way of a heat exchanger (a closed loop).

Depending on the year-round weather conditions and the climate in your area, one of these systems can work very

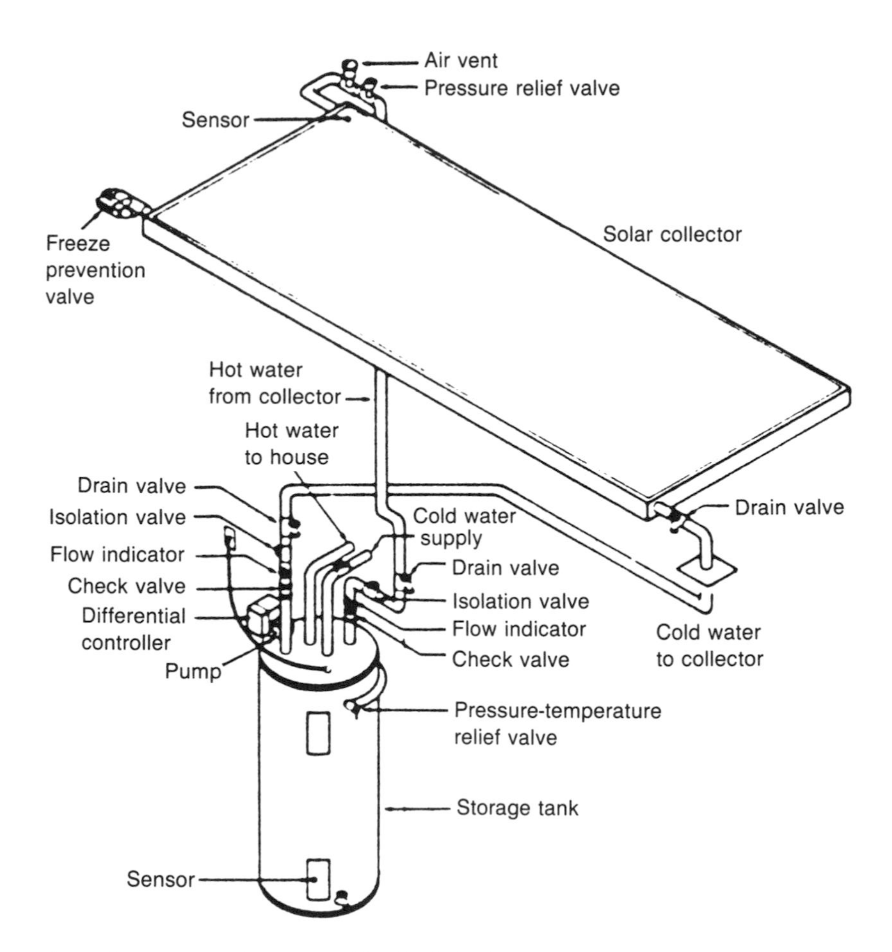

An active pumped system (*Florida Solar Energy Center*).

well. Both these systems are "flat-plate" collector systems, meaning they use the conventional flat solar collectors that have been in use in this country for the past 100 years.

In an active system, a pump circulates water (or a non-freezing fluid) through the collectors, which are usually located on the roof of the house. The water in the collectors is heated by the sun, then returned to the storage tank in the home where it is kept hot until needed. Solar systems typically use better insulated, larger tanks than do electric or other fossil fuel type systems, so the water stays hot longer. This helps assure the availability of hot water during the night and when there are a couple of days of cloudy weather.

In the past, many systems have been damaged by freezing weather that broke pipes or other components in the collectors. During the past few years, however, the industry has resolved most of these problems. Researchers have found that the majority of problems with freezing conditions were caused by improperly installed freeze protection valves or malfunctioning valves or other controls designed to eliminate weather-caused problems.

In the warmer parts of the country, drinkable water is typically circulated through the system and back to the storage tank. As an added precaution against freezing weather, systems in the colder parts of the U.S. often circulate an FDA-approved antifreeze through the system. When this liquid heats up, it passes through a heat exchanger which transfers the heat directly to the drinkable water. Double walls in the heat exchanger and other structural designs assure that the two liquids do not come in direct contact. Another popular freeze-proof design is called the drainback system. Like the antifreeze system, it is a closed-loop system with a heat exchanger. In freezing weather, the water that could cause damage drains back into a small holding tank whenever the pump turns off. Some other types of freeze protection involve draining water out of the system during very cold weather or circulating warm water through the collectors to prevent freezing.

It should be noted that all solar systems include some type of back-up heater for extended periods of bad weather or times of excessive use (like when the in-laws and your cousins all come for a long visit at the same time). When solar systems are put into existing houses, contractors often leave the electric or gas water heater as the back-up system. These units turn on to heat the water when the solar system can't provide enough hot water to meet the family's needs. Tanks built for solar systems, which are used in newer houses, typically include an electric heating element to take care of this need.

Passive Solar Systems

These systems are often chosen in the southern United States because of their simplicity and relatively low cost. Two major types are widely used around the country, and both are highly reliable. The simplest type of system is the Integral Collector System—more popularly called a "Breadbox" or "Batch Water Heater"—in which the collector and the storage tank are combined in one unit. The unit is usually located on the roof or on the ground near the house so that the sun striking the collector goes directly into the storage tank, where it heats the water. The hot water then flows downward into the house.

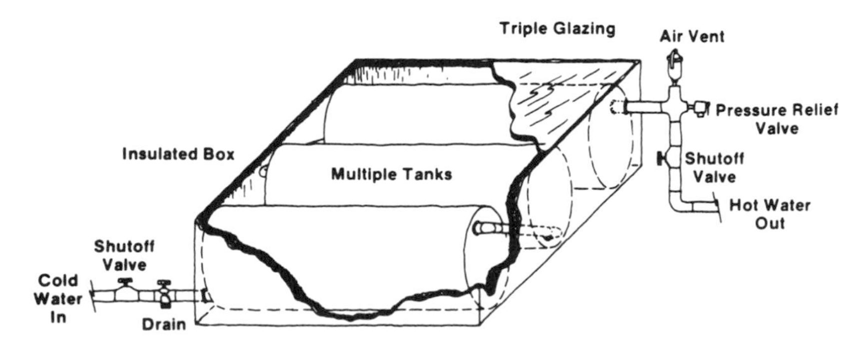

An integral collector storage solar system (*Florida Solar Energy Center*).

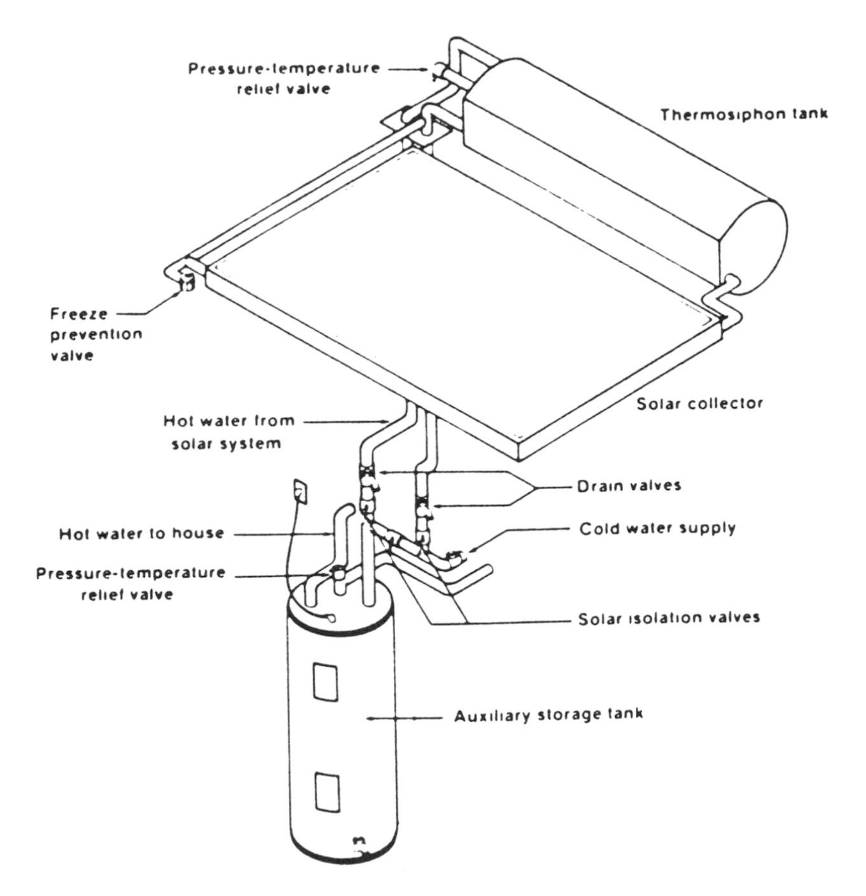

A thermosyphon system (*Florida Solar Energy Center*).

The other type of passive system is a "Thermosyphon" unit, in which a storage tank is located on the roof above the collectors. As the water in the collectors is heated and becomes lighter, it rises naturally into the tank above it. The heavier cold water sinks to the lowest point in the solar system, which is the collector. Like other passive systems, this type is highly reliable because it uses no moving parts. However, these systems can get quite heavy when filled with water, so older homes may require roof reinforcement or other structural changes to the home.

Finally, note that solar dealers in your area might carry other types of active and passive systems, which range from tubes that concentrate sunlight to heat the water to very slim units that look like skylights. In one system, steam is produced by solar troughs which concentrate sunlight onto a tube. The steam carries the heat to the hot water tank where it releases the heat as it is condensed. A couple of gallons of distilled water for drinking is the daily by-product of such a system, providing even greater savings for those of you who regularly buy bottled water. Another new type of system, called a geyser thermosyphon water heater, boils a water/alcohol mixture to raise the fluid and force circulation.

Because there are more than two dozen national manufacturers of solar collectors in the United States today, you will find a wide variation depending on local market conditions. Like anything else, it pays to shop around, not only for price or quality, but also for advice from local contractors who know your area's weather best. Scientists use the term "insolation" to refer to how much sunlight is available in a given area, and the amount of insolation and overall climate conditions can help determine the best type of solar system for your family's needs.

How does a solar system work?

The chief component of an active solar system is the collector. Solar collectors absorb the sun's energy and change it into heat energy.

The typical flat-plate collector, which is the type most widely used in the United States, is a rectangular box. Most collectors are between two and four feet wide, four to twelve feet long, and between four and eight inches thick. At the bottom of the collector is an absorber surface usually made of copper or aluminum and coated black to absorb as much sunlight as possible. A series of fluid tubes run lengthwise through the box, and liquid from the storage tank flows

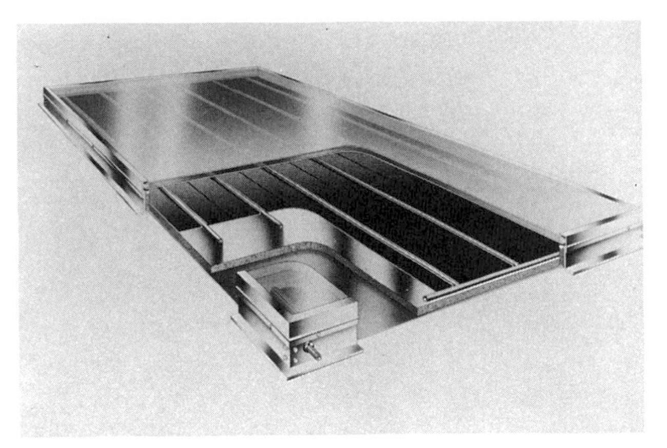

A cutaway look at a flat-plate solar collector. Under the tubes are the absorber plate and insulation (*American Energy Technologies Inc.*).

through the tubes. As the absorber plate is heated by exposure to sunlight, it transfers heat to the liquid in the tubes.

The bottom and sides of the absorber within the collector box are insulated to cut down on heat loss. In addition, a special translucent glass or plastic cover is placed on top of the collector to allow sunlight to strike the absorber. This covering also reduces the amount of heat that can escape from the system.

Solar collectors are usually mounted on the roof of the building, though they can be located on the ground or on another structure nearby. To be most effective, the collectors should be oriented to the south, or within forty-five degrees east or west of due south, and tilted between twenty and forty degrees to horizontal. If the slope of your house roof will not allow the collector to be mounted flush against the roof and still be oriented properly, the collectors can be mounted angled to the roof. Most people prefer having the collectors mounted right onto the roof because it is stronger and looks better aesthetically. However, collectors can be mounted on frames and angled away from the roof without losing any of their effectiveness.

To give the best orientation to the sun, collectors can be ground-mounted if the roof angle is not satisfactory (*American Energy Technologies Inc.*).

For proper solar orientation, collectors can be mounted on rigid frames (*American Energy Technologies Inc.*).

In general, there are four types of solar collectors.

- Unglazed flat-plate liquid collectors are usually used for swimming pool heating systems. Water or another liquid is heated by the sun in a flat collector which does not have a covering.

- Glazed flat-plate liquid collectors are the most commonly used type of solar collector, and are typically used for residential water heating systems. The liquid in the collectors is heated by the sun in a flat collector which has a cover of glass or other transparent material.
- Air-type collectors are used to provide space heating in homes. The sun heats the air in the collectors, and the air is then circulated into the home by a fan.
- Linear tracking concentrator collectors are most often used for commercial and industrial systems where higher temperatures are required, but they can also be used for residential water heating systems. Curved reflectors, called "troughs," concentrate the sun's light onto a liquid within a pipe at the focal point of the collector. The collectors are usually mounted on a tracker to follow the sun to receive the most direct sunlight. Some of these systems may also purify water.

The second major component of a solar system is the storage tank. Solar systems usually use a specially designed and sized, super-insulated tank, though many systems use converted electric tanks or hook up the solar tank with the conventional tank. In a sense, solar systems preheat the water for use in the home. During sunny weather, this is enough to bring the water to the desired temperature, with systems capable of reaching 140 degrees Fahrenheit or more—more than enough for all household needs. During bad weather, the back-up heater boosts the water to the desired temperature.

Active systems that use pumps and controls to circulate the water usually have these two components mounted near or on the storage tank. The pump is a small circulating type, usually 1/100 to 1/12 horsepower. The controller, which regulates when, how long, and sometimes how fast the pump operates, is usually a solid-state electronic device. It senses when the collector is able to heat the water in the tank and

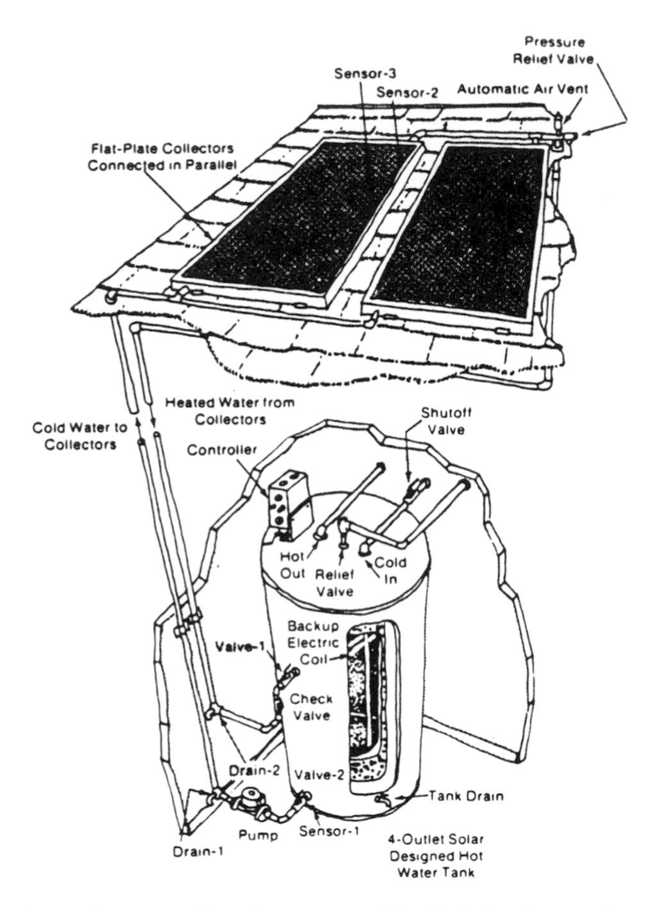

A complete solar water heating system (*Florida Solar Energy Center*).

turns the pump on. When the water in the tank heats to within three to five degrees of the collector temperature, the controller turns the pump off.

Other devices are sometimes used to control the flow of water between the tank and the collector. Timers can be used to operate the pump in some systems. Switches can activate the pump when the collector heats to a specified temperature.

Growing in popularity as a controller is a solar electric (photovoltaic)-powered device, in which sunlight striking the small photovoltaic panel is converted to electricity, directly powering the circulating pump. In this type of system, you don't have to connect the system's pump to your electricity. Since the pump uses only a tiny amount of electricity, this gives one more measure of energy independence and savings. It also reduces the need for controllers, since the pump only works when you need it—when the sun is out!

One other factor needs to be considered when deciding how well a system will work for you. It is important that your roof (or other location where the collectors will be mounted) is not shaded during the hours of 9 A.M. to 3 P.M., when the sun is shining brightest. Such objects as trees, buildings, fences, and other structures can cast shadows during the day which will shade your collectors and cut down on their performance. This problem can be especially difficult to accurately assess before installation because of the sun's changing path through the sky during different seasons.

Fortunately, a number of tools have been designed to help contractors measure the "solar window" of your home and determine any shading problems. Be sure to ask your dealer about using a sun angle chart or other measurement device to assess the potential impact of shading.

The New York State Energy Task Force has put together a simple method you can use to estimate whether shading will cause any problems at your home. They suggest that you stand where the collectors are to be placed and face due south. Point so that your finger and your eye are horizontal,

Table 3-2. Adjustments for estimating the sun angle (*U.S. Department of Energy*).

Latitude	12 o'Clock Position = 0° Bearing		11 o'Clock Position (East), and 1 o'Clock Position (West) = 30° Bearing Angle	
28°N.	18 inches	(47° Alt.)	12 inches	(30° Alt.)
32°N.	14 inches	(34° Alt.)	10 inches	(26° Alt.)
36°N.	12 inches	(30° Alt.)	9 inches	(23° Alt.)
40°N.	10 inches	(27° Alt.)	8 inches	(20° Alt.)
44°N.	9 inches	(23° Alt.)	6 inches	(17° Alt.)
48°N.	8 inches	(20° Alt.)	6 inches	(14° Alt.)

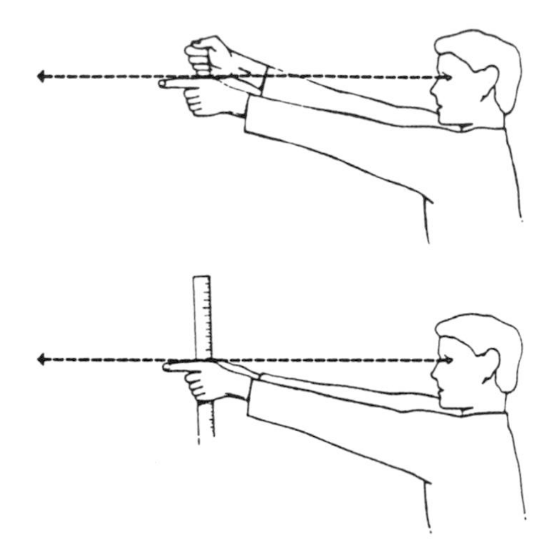

Estimating the solar window (*U.S. Department of Energy*).

and hold a yardstick vertically in your other hand so that the top of the yardstick extends above your horizontal arm by the exact number of inches determined by table 3-2.

You should look over the top of the yardstick at true south and both thirty degrees east and west to determine the shading effects. Basically, any object that is above the top of the yardstick will cast a shadow on the solar collectors, while objects below the top of the yardstick won't cause any shading problems.

How do I know what size system is needed?

One of the reasons why solar energy can be efficiently used to provide hot water for homes is the fact that extremely high temperatures aren't usually needed for residential use. While commercial laundries might need water at 180 degrees Fahrenheit for cleaning clothes, we rarely need water much above 130 degrees Fahrenheit for our needs. Dishwashers and clothes washers usually call for water temperatures at this point to adequately do their job. Otherwise, people could be very content with water not much higher than 105 or 110 degrees Fahrenheit, which is satisfactory for showers and tubs, hand washing, and other uses where water comes in contact with the human body. It's not very easy—or safe—to withstand hot water much higher than that.

To figure out what size system you need for your home, you can begin by thinking about how much hot water your family uses. How many people live in the house? How many loads of laundry are done? Do you use an automatic dishwasher? How many baths or showers are taken every day? The answers to these questions can help calculate the water usage. Table 3-3 gives average daily amounts of hot water usage for these and other activities.

You can also use some rules-of-thumb to estimate your hot-water needs. In general, figure that each adult in your family uses about twenty gallons of hot water every day, and each child uses about fifteen gallons per day. The average family of four, then, uses somewhere around seventy gallons of hot water a day. You then add to this amount the water used by major appliances and other products in your home. Even a fairly new and efficient clothes washer is going to use about thirty gallons of hot water for a standard load when it is set for both hot wash and hot rinse. You can cut this number in half if you use the warm wash/warm rinse setting, but that's still equal to the amount of water used by one of your family members for a full day. Add the water used by your automatic dishwasher, some food preparation in the kitchen, and a couple of other uses, and you'll see that your daily need for hot water is probably more than you thought it was.

This is why properly sizing a system is very important. One of the biggest advantages of solar heating systems is that they are modular. Collectors come in several basic sizes—primarily four feet by eight feet or four feet by ten feet. Water tanks also are available in several standard sizes, such as 82 gallons, 120 gallons, or larger.

Collectors and solar systems sold in the United States are subject to strict performance and exposure tests (see the next

Table 3-3. Domestic hot water consumption (*Solar Energy Industries Association*).

HOT WATER USE (GALLONS)

I. CLOTHESWASHING MACHINE

	Full Loads		
	Compact	Standard	Large
1. 1979 and Later Models			
a. hot wash/hot rinse	22 gal.	30 gal.	38 gal.
b. hot wash/warm rinse	17	23	29
c. hot wash/cold rinse	12	16	20
d. warm wash/warm rinse	11	15	19
e. warm wash/cold rinse	6	8	10
f. cold wash/cold rinse	0	0	0
2. Pre-1979 Models			
a. hot wash/hot rinse	27 gal.	38 gal.	48 gal.
b. hot wash/warm rinse	22	30	39
c. hot wash/cold rinse	15	20	25
d. warm wash/warm rinse	16	22	29
e. warm wash/cold rinse	9	12	15
f. cold wash/cold rinse	0	0	0

II. DISHWASHING

	Small	Large
1. Dishwasher Machine		
a. Pre-1979 Models	10 gal.	14 gal.
b. 1979 and Post-1979 Models	9	10
2. Hand Dishwashing	6 gal. (4-8 gal. range)	

III. PERSONAL HYGIENE

1. Showering	
a. Regular Shower Head	12 (1-3 gal. of hot water per min.)
b. Flow-Restricting Shower Head	6 (0.5-1.5 gal. of hot water per min.)
2. Tub Bathing	20 (10-30 gallon range)
3. Other (Hand-Washing)	4 (2-6 gallon range)

IV. FOOD PREPARATION 3 (2-4 gallon range)

These test stands at the Florida Solar Energy Center show several different types of solar collectors. The instruments in the foreground measure the amount of available sunlight (*Florida Solar Energy Center*).

section). Each solar collector is tested and rated at producing a certain amount of Btus per collector—most in the United States average around 25,000 to 30,000 Btus per system. A solar contractor or dealer can work with you to determine how many collectors are needed to work best for your family's needs.

For example, a family of four in Florida with typical water usage would probably find that one four by ten collector and an 82 gallon storage tank would be sized properly, while a family of three in New York might need two collectors of that size and a 120 gallon tank.

This modularity means that solar systems will work effectively in just about every part of the United States. The systems are sized to meet your family's needs as well as to take into account the local weather factors. Just remember the bottom line—if you've seen the sun shining outdoors, a solar system can provide for a large part of your water heating needs.

Keep in mind that these estimates are just guidelines. You should work with a qualified solar technician or salesperson to determine the exact size system your family needs. While many plumbers, electricians, and other tradespeople are knowledgeable about aspects of solar systems, experience has shown that a specialist in solar energy will be the best source of assistance. Your system will work best if it is properly sized and installed.

For a list of manufacturers, distributors, and designers of solar energy systems contact the Solar Energy Industries Association (SEIA, see chapter 10 for address), and ask for information on ordering a copy of *Solar Thermal: A Directory of the U.S. Solar Industry.* There may also be a SEIA chapter in your state.

How can I pick the best solar system for my needs?

There really is no one system that can be called the best for any particular home. Instead, there are different types of systems offered by various companies. Like anything else, a wise consumer will shop around.

You can make some decisions on the best system by consulting the Solar Rating & Certification Corporation's (SRCC) ratings for solar collectors. SRCC is an independent national, nonprofit rating organization, established by state government energy officials and the solar industry. It certifies solar collectors and entire solar systems, and provides basic information on the Btu output and other general performance characteristics. Many states require that systems be tested by the SRCC before they can be sold in their state.

When buying a solar system, ask your dealer to show you the SRCC sticker that will be affixed to the collector. You can use this information to compare different types of collectors and systems, much like car buyers use the miles-per-gallon information sticker on new vehicles to make comparisons. The dealer can also show you the SRCC rating sheet for collectors and systems which can be used to compare the performance of different collectors. Copies of the current *Directory of SRCC Certified Solar Collector and Water Heating System Ratings* are available from SRCC (address is given in chapter 10).

COLLECTOR RATING NUMBERS

METRIC (SI Units) Megajoules Per Panel Per Day					ENGLISH (Inch-Pound Units) Thousands of Btu Per Panel Per Day			
CATEGORY (Ti-Ta)	CLEAR DAY 23 MJ/m^2.d	MILDLY CLOUDY 17 MJ/m^2.d	CLOUDY DAY 11 MJ/m^2.d		CATEGORY (Ti-Ta)	CLEAR DAY 2000 Btu/ft^2.d	MILDLY CLOUDY 1500 Btu/ft^2.d	CLOUDY DAY 1000 Btu/ft^2.d
A(-5°C)	34	26	17	SOLAR SRCC CERTIFICATION	A(-9°F)	32	24	16
B(5°C)	31	23	15		B(9°F)	30	22	14
C(20°C)	27	19	11	SRCC Standard 100-81	C(36°F)	25	18	10
D(50°C)	17	9	3		D(90°F)	16	9	3
E(80°C)					E(144°F)			

MANUFACTURER: American Energy Technologies
ADDRESS: P.O. Box 1865, Green Cove Springs, FL 32043
EQUIPMENT TYPE: Glazed Flat-Plate Liquid-Type Solar Collector
MODEL #: AE-26 BRAND NAME: American Energy FLUID: Water
CERTIFICATION DATE: April 1985

COLLECTOR SPECIFICATIONS

Gross Area:	2.42 m^2	26.0 ft^2		Net Aperture Area:	2.16 m^2	23.3 ft^2
Dry Weight:	48 kg	105 lb		Fluid Capacity:	4.2 L	1.1 gal
Max. Oper. Pressure:	1138 kPa	165 psig		Max. Opr. Temp:	100 °C	212 °F
Max. Wind Load:	kPa	psf				

Flow Rate at Which Collector Was Tested: 35 mL/s 0.55 gpm
Manufacturer's Max. Recommended Flow Rate: 114 mL/s 1.8 gpm

COLLECTOR MATERIALS

Frame: Anodized Aluminum
Cover (Outer): Low Iron Tmprd Glass Thickness: 3.0 mm 1/8 in. Transmissivity: 0.91
Cover (Inner): N/A Thickness: N/A mm N/A in. Transmissivity:
Absorber Material: Copper
Absorber Coating: Black Chrome over Nickel
Insulation (Side): Foil-Faced Polyisocyanurate
Insulation (Back): Foil-Faced Polyisocyanurate

REMARKS

Thermal performance ratings based on results from Model MSC-21.

TECHNICAL INFORMATION

Efficiency Equation [NOTE: (P) = TI - TA]

				Y Intercept	Slope
SI η =	0.6601 - 2.60 (P)/I - 0.024 (P)^2/I	SI		0.6735	- 4.25 W/m^2.°C
IP η =	0.6601 - 0.4576 (P)/I - 0.0024 (P)^2/I	IP		0.6735	- 0.7489 Btu/hr.ft^2.°F

Incident Angle Modifier [NOTE: (S) = 1/cos θ - 1]

$K\alpha\tau$ = 1.0 -0.207 (S)

SOLAR RATING & CERTIFICATION CORPORATION

Your contractor can show you a rating sheet like this for the collectors you are considering purchasing. This information can be used to compare the Btu output of collectors (*Solar Rating & Certification Corp.*).

Other organizations also test and certify solar collectors and systems. All equipment sold in the state of Florida, for example, must be tested by the Florida Solar Energy Center

(FSEC). A number of states accept FSEC certification for products sold in their states. You can obtain useful information on the performance of many available solar systems by contacting the Florida Solar Energy Center (see chapter 10) for a copy of their free booklet, *Intermediate Temperature Thermal Performance Ratings.* You can use this booklet for general information as well as to work with your solar installer to choose the best system for your needs.

SOLAR COLLECTOR CERTIFICATION

FLORIDA SOLAR ENERGY CENTER
300 STATE ROAD 401
CAPE CANAVERAL, FL 32920

FSEC 1235
MANUFACTURED BY: MODEL #
Manufacturer's Name 6789
Street Address
City, State and Zip Code SERIAL #

has been tested for thermal performance and meets the minimum standards established by the Florida Solar Energy Center as directed by Section 377.705 Florida Statutes.

THERMAL PERFORMANCE RATINGS*

Low Temp. (35°C, 95°F)	27.700 kJ/day	26.300 Btu/day
Intermediate Temp. (50°C, 122°F)	18.800 kJ/day	17.800 Btu/day
High Temp. (100°C, 212°F)	1.300 kJ/day	1.200 Btu/day

*Based on an assumed standard day for Florida.

GROSS COLLECTOR AREA: 2.688 m^2 (28.93 ft^2)
COVER PLATE AREA: 2.601 m^2 (28.00 ft^2)
COLLECTOR LENGTH: 2.188 m (7.18 ft)
COLLECTOR WIDTH: 1.228 m (4.03 ft)
COLLECTOR WEIGHT: 52.3 kg (115.3 lb)
FLUID CAPACITY: 7.9 L (2.09 gal)
REC. FLOW RATE: 1.9 Liquid mL/s (0.5 gpm)
MAX. OPERATING PRESSURE: 827 kPa gauge (120 psig)
MAX. LOAD: WIND 1628 Pa (34 psf)
THERMAL PERFORMANCE EFFICIENCY (ASHRAE 93-77)
Y INTERCEPT: 56.6
SLOPE: 646 $\frac{\text{Watts}}{m^2 \cdot °C}$ (114 $\frac{\text{Btu}}{ft^2 \cdot hr \cdot °F}$)
INCIDENT ANGLE MODIFIER, AXIS 1: 0.05 AXIS 2: N/A
USE RESTRICTIONS: none

The FSEC certification label (*Florida Solar Energy Center*).

Note that ratings are only one criteria that should be used in choosing a solar system. You need to also consider the quality of installation, cost, availability of service and parts, the expected life of the system, and the reputation of the dealer. See the section on what to look for in a system, beginning on page 45, for more information on shopping for the right system.

Are there any tax benefits available to solar system owners?

To help stimulate the introduction of solar water heating systems following the energy crises of the 1970s, the federal government instituted *residential* tax credits in 1977. This program reduced the cost of a solar system by 40 percent, and was instrumental in building the solar industry to a peak in 1985 of several hundred companies and approximately 30,000 employees, producing 20 million square feet of solar collector area. At the end of that year, however, the credits expired. However, if you own a business, you can still receive a 10 percent commercial federal tax credit, at least through December 1991, and probably longer.

Many state governments had instituted their own tax credit and incentive programs during the 1970s, and a number of the states have kept their own tax incentive programs in effect. In 1989, about half of the states continued to offer tax credits or other financial incentives, including property tax and sales tax exemptions, to homeowners and to businesses purchasing solar water heating or photovoltaic electric systems. The following list covers the state incentive programs that were in effect in 1990 for residential solar energy systems. Consult your local state tax office for information on current programs in your state. Several other states are considering tax credits starting in 1991.

Connecticut: Solar equipment is exempt from sales and use tax, and property tax exemptions are available in many towns.

Florida: Renewable energy source exemptions for real property upon which a solar system is installed, not to exceed 8 percent of the assessed value of the property.

Hawaii: A 35 percent state tax credit, in effect through December 1992.

Idaho: A tax deduction of 40 percent of the cost of a system for the first year, and 20 percent for up to three years, with a maximum of $5,000.

Indiana: Property tax deductions are administered locally. The value of the solar system is deducted from the appraised value of the house.

Maryland: Counties have the option to offer property tax exemptions.

Michigan: An income tax credit of 30 percent for the first $2,000 of cost, and 15 percent of the next $3,000. A maximum credit of $1,050 is available for single family homes.

Montana: An income tax credit of up to $125 for solar systems, with 5 percent of the first $1,000 of system cost and 2½ percent of the next $3,000 qualifying for the credit. Owners of solar systems are eligible for a ten-year property tax exemption based on the value of the system. For owners of single family homes, property taxes are exempted on a maximum of $20,000 of the system's appraised value.

Nevada: Property tax credits.

New Hampshire: Towns may give property tax exemptions at their option.

New Jersey: Solar systems are exempt from state sales and use taxes. No construction permit fees are required for the installation of solar equipment.

New York: A real property tax exemption, exempting property owners from paying increased property taxes for fifteen years after the system has been installed. Systems that meet a fifteen-year payback are also eligible for low-interest loans.

North Carolina: A 25 percent tax credit.

North Dakota:	An income tax credit and a property tax credit of 5 percent per year for three years, based on the cost and installation of the system. Property tax exemptions are in effect for five years following the installation of the system.
Texas:	A property tax exemption.
Virginia:	Some localities offer property tax exemptions.

What do I look for when buying a system?

If you talk with qualified solar contractors about the different types of systems sold in your area, you will find a great deal of information to use in evaluating the best type of water heating system for your needs. You should also contact some of the organizations listed in chapter 10 of this book for additional consumer information.

Make a list of comparisons among your choices regarding the background of the contractor, the quality of the product,

Solar water heating can work efficiently in all climates and in all parts of the country (*American Energy Technologies Inc.*).

the cost based on a written estimate, and the details of the contract with a written warranty. Keep in mind that you'll get best performance from solar components when they work together well. This means that you are usually better off buying a single package from a company, rather than component parts that are not specially matched. If your dealer does not sell complete systems, be sure that he has ample experience in selecting compatible components.

Helpful information from Northeast Utilities can help you make the right choice. This New England utility company operates a program called "Operation SOLAR," designed to help their customers use solar energy appropriately. They suggest that when shopping for a system, you look for a durable collector, a proven system that has been sold in your area, efficiency tests from SRCC, performance predictions from the dealer, specific warranty information, maintenance details, an owner's manual to help you take care of the system, monitoring devices such as temperature or pressure gauges to let you know everything is working, a commitment from the manufacturer to support the system, and a good price.

When choosing a contractor, ask about his solar installation experience, names of satisfied customers, required permits and building regulations, warranty and service details, available financing programs, and other information to show that the dealer is experienced and will provide service as needed.

You should call several solar contractors listed in the Yellow Pages of your local phone book. See if there is a listing for a local affiliated chapter of the Solar Energy Industries Associations (see list of current chapters in Chapter 10). Talk with them about price, efficiency, service options and warranties. Ask them to provide the names of customers who have bought from them in the past. Many contractors will provide pictures of their installations. Call these customers and find out how they like their systems, how well the systems are working, and what kind of energy savings they have had.

The warranty you get with your system can be very important, so ask about the details before signing the purchase agreement. If a component fails because of improper installation, who will be responsible for the repair? Many installers offer at least a one-year warranty, along with whatever warranty coverage the manufacturer provides (many cover material and workmanship on collectors for five full years). You should also check with your homeowners' insurance agent to determine if you are covered for freeze damage that may fall outside of the warranty provisions.

Members of the solar industry follow a Code of Ethics set forth by the Solar Energy Industries Association, the national trade organization serving the manufacturers, distributors, component suppliers, and contractors in the field. SEIA works with its member companies through training and education programs, product development, and other business programs designed to give consumers the best quality products to meet their needs. Also, many local government officials enforce strict building codes and standards that assure consumers that products like solar water heating systems meet their standards. Many communities in the United States also require specific licenses for contractors who install and maintain solar equipment.

What happens to solar systems in freezing weather?

As noted earlier, an important part of a solar water heating system is freeze protection. In some active systems, the controller can be designed to turn the pump on in near-freezing weather, running warmed water through the collector. This is called a draindown system.

Another type of system uses manual or automatic draining of the collector to prevent the pipes from freezing. In this drainback system, water from the collectors empties into a holding tank when the temperature drops near freezing.

A freeze valve can also be used to protect the system. When temperatures get near the freezing point, the valve opens, and water pressure forces water through the collectors and out through the valve.

The previously mentioned "closed-loop" system circulates an antifreeze solution, usually glycol, through the collectors rather than water. This liquid, usually glycol, never freezes, and is circulated through the water tank where the heat is transferred to the water via a heat exchanger.

Freeze protection should be addressed even if you live in a warm climate where the weather rarely gets cold. It pays to be prepared for the occasional freeze.

Do I have to change my living habits to use a solar water heater?

No. Because of the back-up element, you will always have enough hot water for your needs regardless of your usage or the weather conditions. However, there are some things you can do to improve the efficiency of your system and cut down its reliance on the back-up element.

First of all, you can get the most benefit from your system by using most of the hot water in the late morning and early afternoon. Obviously, this is when the sun shines the most, so you will be using almost totally solar-heated water at these times.

Another suggestion is to spread out heavy hot water usage during the week so that you don't use all of the solar-heated water at one time. Do a load of laundry today and one tomorrow instead of both today. Use of a water-saving shower head will save even more energy and money on your water bills.

In general, though, using solar energy doesn't call for sacrifices or compromises. If you want to make some minor lifestyle adjustments, you can save even more energy. If not, at times you'll just use more of your back-up power to heat water for your home.

Will a solar water heating system give me all the hot water I need?

Probably not. But it will provide somewhere between 60 and 80 percent of your hot water needs. This is because you'll certainly encounter some conditions which call for your back-up element to be used. These include extended periods of cloudy weather (usually three days or longer), times when you have house guests who increase your typical hot water usage, and other times when the system just can't provide all of the hot water. In Arizona, for example, you may be able to turn off the back-up element in the water heater from March through October and let the system provide 100 percent of your hot water during that time period. In Wisconsin, though, you may need the conventional heater element available during ten months of the year, so you're getting part of your hot water from the back-up system.

Using the back-up element ensures that you will always have hot water. But because the average solar system meets more than three-fourths of your annual hot water needs, the savings can be significant.

Can I build my own solar system?

Sure you can, if you have the time, the experience, the knowledge, and the willingness to maintain and repair the system yourself. The question is why would you want to do this?

As the solar industry has matured, many educational programs and ongoing training efforts have qualified hundreds of solar contractors around the country to properly install solar water heating systems. The result is that systems usually work to their expectations, and can be maintained with minimal annual effort. The cost of the installation is actually fairly low, and buying a commercial system with an SRCC or another rating means that it has successfully undergone rigorous testing for quality, durability, and performance.

Many homeowners have built their own batch water heating system, which is fairly simple to construct and install. However, if you're not experienced with construction, roofing, electrical, and plumbing skills, it's probably not worth the effort. You also will need detailed engineering information, plans, and other materials to help you properly design, size, and install your own system. Of course, the same skills would also allow you to build your own refrigerator, but we wouldn't recommend that either. Give it some serious thought before attempting the job yourself.

Summary

Homeowners often ask if they can afford to buy a solar water heating system. The answer is very simple. If you're paying a utility bill every month, you can afford to buy a solar system. The energy savings will pay back your investment, and you will have many years of nearly free hot water for all your family needs.

Even if you're not planning to stay in your present home for more than a few years, you start getting an immediate payback on solar water heating systems from the monthly energy savings. This is especially true in new construction, where the cost of the system can be added to the mortgage

Solar water heating can give you many years of free hot water (*FAFCO*).

payments, and savings on monthly utility bills more than offset the slightly increased mortgage payments.

On the other hand, you can take comfort in the fact that systems are designed to last a long time. Most manufacturers expect their solar components to give twenty years or more of reliable service with only minimum maintenance. Oftentimes, the systems last much longer. A couple of years ago, a developer in Daytona Beach, Florida, tore down some older homes to make room for a new housing development. One of the homes was especially interesting. Over the years a number of improvements and changes had been made to the house, but the residents still continued to use solar energy to provide hot water—water heated by rooftop solar collectors that had been installed in 1935! Not all systems will work for fifty years, of course, but solar energy equipment has generally been characterized by well-made products that can last a very long time.

In the Florida survey mentioned earlier in this chapter, two-thirds of the homeowners interviewed said they "seldom" or "never" have any problems with their solar water heating systems. Seven percent said they did have problems, but 88 percent of the people who have had problems said that their systems had been fixed.

Finally, if you're planning to build a new home, talk with your mortgage company about energy-efficient mortgages. This new concept is rapidly gaining widespread acceptance around the country, and many major federal mortgage programs are now allowing "credit" for energy efficiency when calculating mortgage qualifications. This means that many mortgage companies are giving larger debt-to-income ratios so that homeowners can afford a larger or better home with bigger monthly payments since their energy costs will be lower. Many states are now putting together energy-efficient mortgage programs for home buyers who include solar water heating and other energy-conserving features in their new homes.

Take a look out your window right now. If the sun is shining, just think about all that free energy falling around you. If the sun's not out right now, keep in mind that the water tank will still be storing hot water heated by the sun, ready for you to use when you want it.

Think about that the next time your teen-ager takes another half-hour shower. Wouldn't it be nice if all that hot water were free?

CHAPTER FOUR

Solar Pool Heating

Enjoying your swimming pool several months longer every year

Do you have a swimming pool in your back yard?

If so, then you really ought to be taking advantage of the most economically attractive use of solar energy available today. Solar pool heating can extend your swimming season several months each year and make that investment in your pool pay off even more with more use and enjoyment. More than 250,000 solar pool heating systems have been installed in the United States, providing efficient heating to residential and commercial swimming pools around the country.

Remember how you felt when you decided to have that pool built, or when you moved into the house that already had a swimming pool? You probably imagined many enjoyable months of the family swimming together and lounging around the pool. The evening swim parties. The kids and their friends spending afternoons outside your home. The great exercise and physical benefits of swimming.

Solar pool heating can extend your swimming season several months each year (*FAFCO*).

But in reality, you probably learned that first year what most pool owners around the United States already knew: The swimming season is actually a very short one, restricted to the summer months in most of the country. A pool's temperature cycle varies with climate and geography so that a three- to four-month swimming season is fairly typical, even in many of the southern states. During the rest of the year, that pool just sits out there, a reminder of how little you probably actually used it when you had the chance.

Sure, fossil fuel heaters are available for swimming pools, and many of them were very popular until the energy crises of the 1970s shot fuel prices sky-high, forcing many pool owners to disconnect or remove their heaters. Today, the annual cost of heating a pool with electricity, fuel oil, or propane heaters can easily range from $500 to more than $2,000 for a few months more of swimming time.

Solar collectors for pool heating systems are usually mounted on the roof of the home. The pool's filtration pump powers the system (*Heliocol U.S.A.*).

That's why the fastest-growing use of solar energy in the U.S. today is for heating swimming pools. While the average system costs around $4,200, the significant savings over the cost of using conventional fuels to heat the pool will pay back that investment in only about three years. Depending on where you live in the U.S., the solar pool heater will extend your swimming season from two to four months, allowing you to use your pool for anywhere from five to eight months (or more) every year in most parts of the country.

Why are solar pool heaters so cost-effective?

The economics of pool heating are extremely positive for a number of key reasons.

First, you really don't need to raise the water temperature

very much to make a big difference in the comfort level. Just before or just after your typical swimming season, the water in your pool might be somewhere between sixty-five and seventy degrees Fahrenheit. If that water temperature can be raised as little as eight to ten degrees, it can extend the swimming season for a few months. Compare this to the fifty to sixty degrees increase needed to heat water for use in your home.

Because of the much smaller differential involved in pool heating, very efficient solar collectors can be used. One of the reasons these collectors are so efficient is because of the relatively low temperatures involved. Scientists already know that the hotter solar collectors get, the more energy they lose to the outdoors, cutting down on their efficiency. But solar pool collectors don't get that hot, thus working efficiently and keeping more heat for the system.

Solar pool heating equipment is a lot simpler, too. You don't need a storage tank like your home's water heater; the pool itself serves as the storage system. And in most cases, the pool's existing filtration pump can also be used to pump water through the solar collectors, cutting down on the need for extra mechanical equipment.

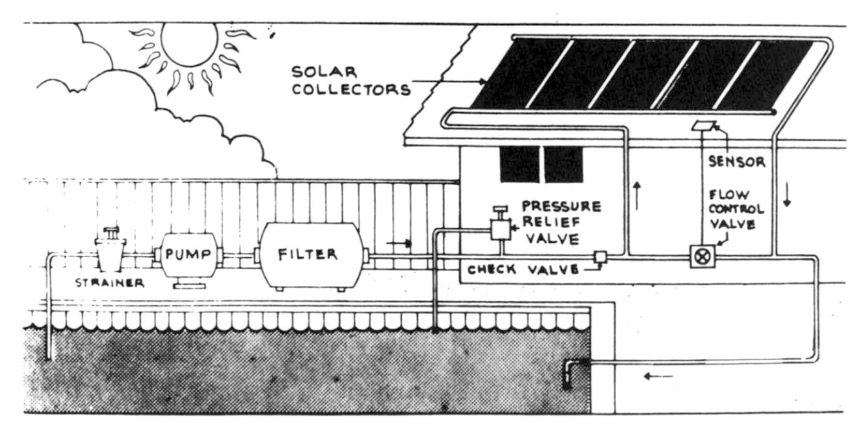

A basic solar pool heating system (*Florida Solar Energy Center*).

You also need to keep in mind that the most use of swimming pools takes place during the daytime, when the sun is shining. You don't need a back-up system, since the water will stay hot enough for use that night and then be reheated again the next day.

Obviously, the biggest reason why solar pool collectors are so economical is because of the high cost of alternative fuels. Because heating a pool is a luxury—not as essential as heating water for indoor household use—the cost of fuel for pool heating is strictly an extra addition to the family budget. Solar heat is absolutely free.

Solar collectors can be mounted to any type of roof. These collectors are on a barrel tile roof in south Florida (*Heliocol U.S.A.*).

How does a pool heating system work?

It's actually a very simple process with only solar collectors and some piping needed to make the system. This means no storage tank, usually no pump, and no other mechanical equipment.

Though a variety of different types of solar collectors have been designed for use with swimming pools, most are made of black plastic material with tubes running through them. The collectors in a pool heating system are not covered with glass or plastic because they are used when both solar radiation and outdoor temperatures are relatively high.

When the sun is shining, water from the pool is pumped through a pipe to the collectors and back to the pool. The sun heats the water as it flows through the collectors.

Most pool heating collectors use ultraviolet screening materials to protect the plastic materials, resulting in a life span of ten to fifteen years or more.

How do I know what size system is needed?

Working with a qualified solar contractor will assure that you get the right size and type of system for your needs. However, you can estimate in advance with some general rules for sizing solar pool heating systems.

Basically, the larger the area of the solar collectors, the greater the temperature rise in the pool water. The best way to estimate is to figure that the area of the solar collectors should be at least 50 to 75 percent of the pool's surface area. This ought to increase the temperature of the water up to ten degrees. For example, if your pool is 20 feet by 30 feet (600 square feet), it will take between 300 and 450 square feet of solar collectors, depending on your climate, to obtain the eight to ten degree temperature rise desired.

In most parts of the country, an increase of ten degrees means the swimming season can be extended a few more months. In Southern California, this increase would allow homeowners to keep the water in their pools at around eighty-two degrees from April through October. Using natural gas to heat a pool during this time period would run around $1,500 per year at current rates. Meanwhile, a pool owner in Florida who followed this guideline would have a swimming season from March through November, with a savings of about $1,200 in natural gas heating costs.

If you want to raise the temperature even higher, just add to the size of the solar collector area. Use double the pool surface area and you can raise the temperature of the water by as much as fourteen degrees.

Also keep in mind that collectors should face south and be mounted at an angle equal to the local latitude (or latitude plus fifteen degrees to maximize winter heating) if possible. Your contractor might have to vary from these guidelines if the location of the house or other factors make these locations impossible, with only minimal loss of effectiveness.

Can I install a solar pool heater myself?

This is a frequently asked question because much of the basic equipment, including the pump and a great deal of the pipe work, is already in place. However, unless you are experienced with both plumbing and electrical work, it is advisable to have a solar contractor do the work. Experience has shown that a job taking six hours by an experienced crew will take two do-it-yourselfers a couple of days to complete.

Many homeowners have tried a do-it-yourself version of a pool heater with less than adequate results. Because a solar collector is basically just a hose or tube filled with water that is heated by the sun, we've heard of examples where a home-

A do-it-yourself approach to solar pool heating—not recommended by the experts! (*Florida Solar Energy Center*).

owner has bought a couple of dozen garden hoses, nailed them to the roof, and pumped the pool water through them. Even though the look of such a system is usually enough to discourage someone from trying this, there's a more practical reason why this isn't very cost-effective. Researchers have estimated that it would take at least *two miles* of half-inch garden hose to raise the temperature enough to make the pool water swimmable. Now you may have a pretty big roof on your house, but two miles of hose?

Can solar energy be used to heat a spa or hot tub?

Spas and hot tubs both pose special problems for energy-conscious owners because of the high temperatures that are needed and the nighttime energy use patterns. Both of these factors reduce the general effectiveness of solar energy systems, but options exist to make them effective.

Solar collectors can efficiently heat water for spas and hot tubs (*American Energy Technologies Inc.*).

Solar collectors can be used to bring the water temperature to 105 degrees Fahrenheit during the day. While the spa or tub is used, the water temperature will drop 1 or 2 degrees every fifteen minutes, which is adequate for most people. However, if the 105 degree temperature must be maintained, you can use a gas, oil, or electric back-up heater. Some homeowners use heat pumps to absorb heat from the air and dissipate it at an elevated temperature to the spa water.

What seems to work best for general use is a solar water heater to bring the temperature to the desired level, and an insulating cover to cut down on evaporation losses when the tub or spa is not in use.

Most people only use their hot tubs for an hour or so each day, leaving plenty of time for the unit to be covered. This combination of solar energy and a cover ought to provide for your heating needs at a much lower cost than using fossil fuels.

What can we do to make the pool system as efficient as possible?

The best way to enhance the solar system's performance is to use a pool cover to help keep the heat from escaping. It has been said that trying to keep pool water warm without a cover is like heating a house without a roof. Most experts recommend a light-colored or transparent cover which will allow sunlight into the pool, adding to the solar performance, while actually absorbing heat and helping raise the temperature. Generally, keeping a pool covered for twelve hours a day can result in a five degree temperature increase in the pool. Covering it for twenty hours a day can raise the temperature ten degrees higher than it would have been otherwise.

A cover will also help keep the pool clean, cutting down on the cost of chemicals and filter maintenance.

Other things you can do include planting bushes and trees or building solid fences alongside the pool (not on the south side) to cut down on heat loss caused by the wind blowing across the pool.

There are also two things that you shouldn't do to keep your pool water warm if you want to reduce your energy usage. First, don't cover the pool area with screening, or you can lose as much as 40 percent of the sunlight that strikes the area. Second, don't paint your pool black in hopes of absorbing more sunlight. You'll only get about 10 percent more solar energy than you would with a lighter color, but you'll have additional maintenance problems since the chlorine in the water will bleach darker colors. Sunlight hitting the pool scatters and reflects throughout the pool, and is distributed throughout the water, so a dark color won't necessarily do a better job of attracting heat.

Other tips

Recent research has found another way for pool owners to save some money while helping utility companies cut down on their peak power usage.

Depending on how long your pool pump runs every day, you could be spending anywhere from twenty to fifty dollars each month on electricity for the pump. If you've had your pool for a while, chances are good that the pump is oversized, not very efficient, and is usually set to run for far too long each day.

Solar systems can be designed to keep the water warm in all types and sizes of pools (*Florida Solar Energy Center Library*).

A way to save energy, then, is to set the pump to run as little as necessary—enough to mix the chemicals throughout the pool and keep it free of debris by drawing water out through the skimmer, floor vacuum, and filter, but not so long that it wastes energy. A study by researchers at Florida Atlantic University in Boca Raton, Florida, found that most people were happy with the cleanliness of their pool when the pump ran for only two to four hours a day—far less than most pool pumps are set to operate.

The way to maximize energy savings is to set the pump's time clock to run two times a day for an hour and a half each time, or three times a day for an hour each time. This should be enough to keep the chemicals mixed and the pool surface free of floating debris. If you time it right, you can operate the pump when the utility company is not trying to meet peak power demand, such as late in the afternoon. This will help keep your electricity costs lower.

Researchers found that reduced pump circulation time doesn't increase the chances of having a dirty pool. As long as water is circulating when chemicals are added, the water and chemicals should be adequately mixed. They also found that algae growth is not reduced by high circulation rates.

The following guidelines can help give you the most energy-efficient pump running times:

- one cycle per day: 11:00 A.M. to 2:00 P.M.
- two cycles per day: 9:00 A.M. to 10:30 A.M., and 9:00 P.M. to 10:30 P.M.
- three cycles per day: 9:00 A.M. to 10 A.M.; 1:00 P.M. to 2:00 P.M.; 9:00 P.M. to 10:00 P.M.

We strongly recommend that you watch your pool closely for the first week or two after you make any changes in the settings. Since so many factors affect the exact amount of time needed for your pool, including shading, amount of nearby debris, climate, etc., you will probably need to experiment with one of the above cycles to find the best time setting for your pump. The bottom line, though, is that reduced pump running time can cut about 60 percent of the electricity you're using now.

The guidelines listed in this chapter can greatly increase your enjoyment of your swimming pool while cutting down dramatically on energy use. Installing a solar pool heater, using a pool cover, and cutting down on pump running times will maximize your use of the pool and greatly extend your swimming season. That means a lot less time during the year when you look at your pool and wish the water was warm enough for swimming.

CHAPTER FIVE

Solar Electricity

Plugging into the sun for electrical power

Imagine a dinner party at your home tonight. Your guests drive slowly down the street, looking for the addresses on the homes. Most addresses are difficult to read from the road, but yours is a pleasant surprise. The house numbers are lit up and easily readable from the street. And you're not using any electricity for that light, since the power was provided by the sun.

Solar-powered house number lights are just one of the many types of photovoltaic (PV) consumer products widely available to the average consumer today. Products like these are low-cost, easy to use, and highly reliable.

Almost all of the solar-charged products and appliances on the market con-

Solar energy can light up your house numbers (*Solarex Corp.*).

tain a battery or capacitor which stores the solar-generated electricity. These batteries slowly take in the electricity (a process known as trickle-charging) and usually hold the charge longer than batteries which are charged by conventional electricity. The popular conventional rechargeable

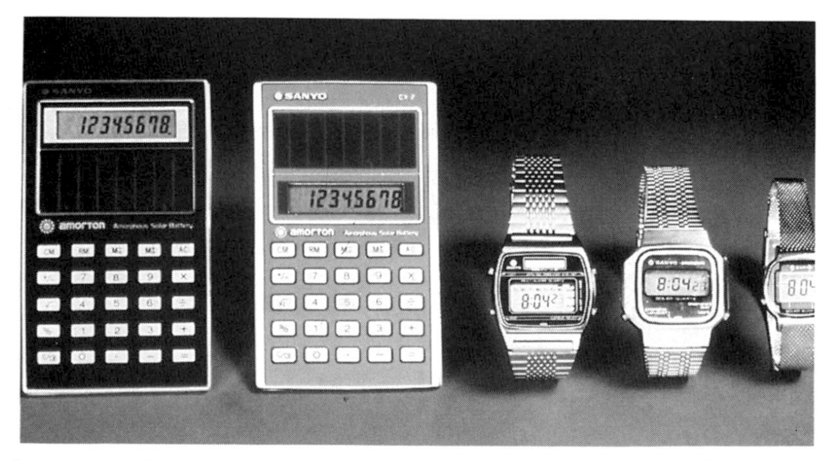

Low-cost PV-powered consumer products like these are very popular today (*Photovoltaic Energy Systems, Inc.*).

nickel-cadmium batteries develop a "memory" problem—in which they often hold less than a full charge—due to high-voltage charging that solar-charging does not create.

All solar appliances work off solar cells, known as photovoltaics, which convert sunlight directly to electricity. The cells can be of a size that covers barely one square inch, such as those used to power a solar calculator, or the solar cells can be joined together into a much larger module, ranging from a few square inches to a few square feet to power lights, water pumps, or fence chargers. The larger the area of the solar cells, the more electricity they will produce.

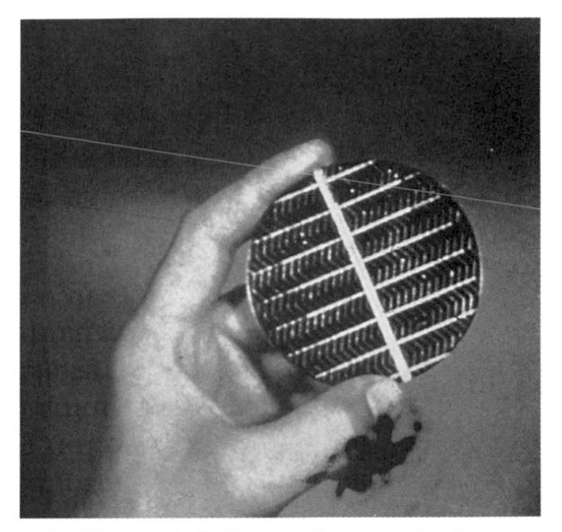

A simple solar cell (*Photovoltaic Energy Systems, Inc.*).

What are solar cells?

Solar cells are materials that absorb sunlight and convert it into electricity. Today's photovoltaics industry is a product of 1950s solid-state electronics, and was greatly stimulated by the need to find power sources for America's space satellites. Its beginnings, though, can be traced back to 1839 when

Different types of materials and different sizes of modules are used in various solar electric systems (*Siemens Solar*).

French scientist Edmund Becquerel observed that light falling on certain materials created a difference in electric response—in other words, it created electricity. He further discovered that the amount of electricity produced varied with the amount of light and intensity. This is called the Photovoltaic Effect—"photo" for light and "voltaic" for voltage.

Solar cells are very thin (about 1/100 of an inch thick) rectangular or circular wafers typically made of silicon, the same basic material that makes up sand at the beach and is used in most of the electronic components in your stereo and other electronic appliances. The silicon can be "grown" in crystals that are interconnected and layered under glass or plastic, or the silicon can be gasified and layered under glass or plastic as a thin film.

A variety of semiconductor types can be used to make photovoltaics, including single and polycrystalline silicon, thin-film amorphous silicon, and several advanced technology materials. Because silicon is so abundant—it comprises more than one-fourth of the earth's crust—and it has the proper physical attributes for the process, it is the basic material for solar cells.

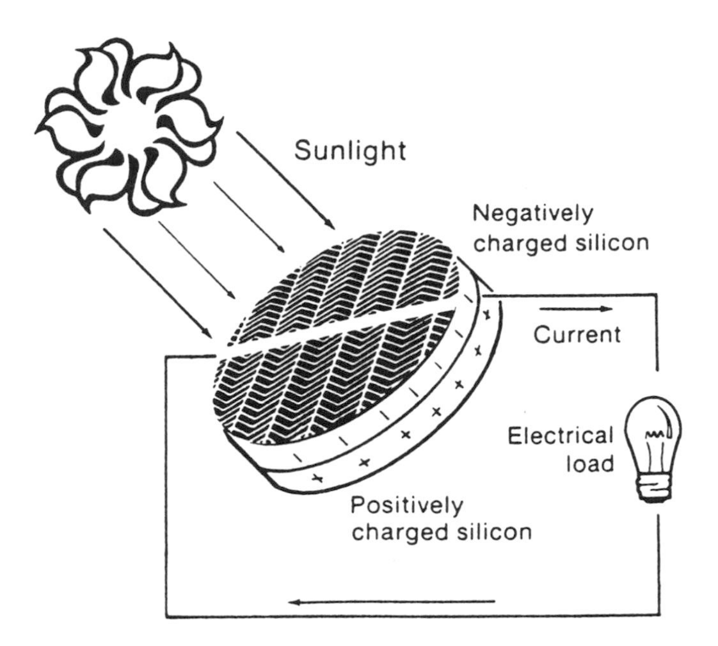

PV cells convert sunlight directly into electricity (*Florida Solar Energy Center*).

How do solar cells work?

No matter what type of manufacturing process is used to make the different types of photovoltaics, when sunlight hits the solar material, the electrons are released. The electrons then flow onto the wires, forming direct current (DC), the same kind of current that flows from a regular battery. A four-inch silicon cell can produce about one watt of DC electricity.

A number of cells (usually twenty or more) or a photovoltaic film can be mounted within a frame under a transparent glass or plastic covering to form a module, and modules can be connected to other modules to form an array. The larger the area of photovoltaics in each module, or the more arrays used, the more electrical power you will have.

You can use the DC power directly to operate many devices, or you can convert it to alternating current (AC) for standard household appliances. Both DC and AC power can be used as it is produced, or can be stored in batteries for later use.

At the present time, solar cells are between 12 and 20 percent efficient, and solar thin-films are 4 to 11 percent efficient. This means that they convert that much sunlight into electricity. Newer cells have been tested at much higher efficiencies, and it is anticipated that they will be available on the market in the near future.

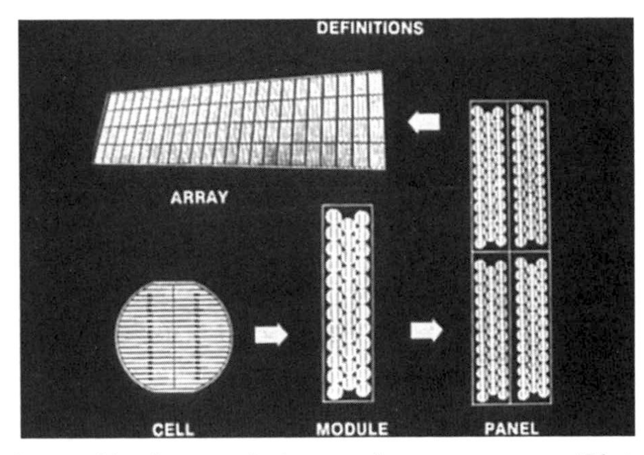

Cells can be combined as needed to produce more power (*Photovoltaic Energy Systems, Inc.*).

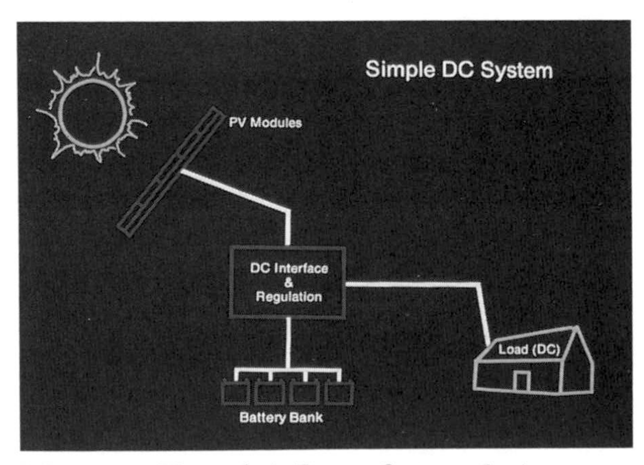

A simple DC system (*Photovoltaic Energy Systems, Inc.*).

An auto battery charger (*Solarex Corp.*).

Battery chargers

One of the most popular consumer products today is the solar battery charger. The most common solar charger uses a four-inch-by-eight-inch solar panel which plugs directly into your automobile's cigarette lighter (you shouldn't be smoking anyway, so here's a practical use for the lighter) and trickle-charges your battery. The small solar panel is placed on the dashboard, and when the car or truck is parked outside, the PV panel converts the sunlight to electricity, providing either six or twelve volt DC current to your battery. Among the biggest users of these solar battery chargers are National Guard units, whose vehicles sit for as long as a month between exercises. In the past, reservists often had to spend most of a day charging truck batteries before they could begin their training. Now the trucks and other vehicles are ready to use when they are needed.

Solar battery chargers can keep your battery fully charged, extending its life. They will also add power during cold weather when the battery charge declines. They also keep your battery topped-off during periods when the car is not being used for a while, such as sitting in airport parking lots or on blocks in your yard during extended periods of nonuse.

PV battery chargers can be left on the dashboard and plugged into the lighter to keep the battery fully charged (*Solarex Corp.*).

There's a built-in reliability measure, too. Because of the natural impedance created by trickle-charging a battery, a solar unit cannot overcharge and damage your battery.

Solar battery chargers are also widely used to charge batteries for boats and other marine applications. In these cases, the solar panel is much larger than that used in automobiles, and is sized to the battery or batteries. You need to make sure when buying a solar battery charging system that it specifies the charge and compatible batteries to ensure maximum efficiency. Because the solar cells are encapsulated between glass or plastic, they are usually impervious to water. With specially-packaged marine solar charger systems, the solar panel and control equipment are connected directly to the boat battery and have plastic-covered connectors to prevent shorting due to exposure with the water.

Lighting systems

On the home front, photovoltaic lighting systems can be very economical. By using solar lighting systems, you will

A PV-powered patio light (*Siemens Solar*).

A PV-powered walk light (*Solarex Corp.*).

save the expense of running electric wires though your home or over/under the ground in your yard. Solar is an ideal power source for providing light on the porch or patio, particularly if you need light for only several hours per night.

When determining whether a solar lighting system is right for your purposes, you have to make sure that no trees or buildings block the sunlight to the solar module. Usually placed atop a lightpole, the module must have solar access throughout the day at almost any angle, so it is imperative that the panel face southward without anything impeding the sunlight during peak daytime hours (9 A.M. to 3 P.M.).

The economics of solar lighting systems are great, since solar lights usually pay for themselves within a year when you take into consideration the costs of running or trenching wires and the time of an electrician. Plus installation time itself takes only minutes. In addition, solar lights use lower-voltage bulbs with higher quality reflectors in the lamp. This

The twenty-one-watt PV panel next to the home's deck will power four lights like the one shown here. Each light will provide light equal to the output of a seventy-five-watt incandescent bulb. A complete system with PV panel, four lights, and all components sells for under $500. Each light will run for more than six hours a night (*Applied Energy Technology Ltd.*).

conserves precious energy and ensures that the same amount of lumens (the standard measure of light intensity) is emitted when compared to conventional lights. Lower-voltage bulbs also last longer because of the smaller amount of electricity running through them, so maintenance costs will be less. Finally, solar lighting is safer, an important feature especially when you have pets or small children at home. First, there are no electrical wires coming from outlets or the lighting units. Even if one of the wires from the solar cells to the light fixture is cut or chewed, the voltage is so low that it will barely be felt and certainly not cause any harm. And because there are no wires, lights can easily be moved to new locations.

Solar lighting systems are available in a wide choice of sizes and light outputs. Larger solar street lights are available, which meet most state and local light-emitting standards. Smaller porch, deck, and patio lights are more common, and

Smaller PV-powered lights are used to mark paths and stairways at night (*Siemens Solar*).

are made by several manufacturers. Finally, you can buy small pathway lights which have very low-voltage lamps that put out a surprising amount of light. The units are mounted on small garden stakes that can be stuck in the ground along a walk or driveway. A sensor turns these lights on at dusk, and they will remain lighted for several hours after dark.

When buying a lighting system, look carefully at two conditions of operation. One is the amount of sunlight needed to completely charge the batteries. While the batteries will receive a charge even on cloudy days, the packaging or instructions should tell how long the lights will stay on at night following bright days and cloudy days. Second, see what the amount of storage capacity is, which indicates the number of cloudy days the light can still be lit without a successive charge of sunlight.

Another very important factor in your purchase decision is the amount of lumens that the solar device emits. In most cases, solar devices will indicate the light output of a standard 15, 60, and 100 watt bulb. This will give you an idea of how bright the light will be. This information will help you decide if a particular light is bright enough to let you see clearly at night on your patio, or just illuminate the path through your yard.

Photovoltaics can be used very economically. Solar lighting is used extensively by the military to provide portable lighting systems in remote areas. A popular consumer item which is available in many hardware stores and consumer electronic and gadget catalogues is the solar security light, in which a solar-charged light is mounted on the side of the home, usually under the eaves or below the gutters. When someone approaches the house after dark, the bright spotlight automatically turns on to highlight the intruder. Solar-charged security lights at a home or business provide safety without the hassle of a complicated hookup, making the installation a simple job for any do-it-yourselfer. Here again, you don't need to run any wires or do any electrical work. You just hook the unit to the outside of your house and position the solar panel so that it faces toward the south.

Motion detectors automatically turn on the PV-powered light when someone enters the area. An electrically-powered security light costs about $150 for the light, bulbs, and electrician to install it. That's about 15 percent *more* than the cost of the PV light—which has no installation costs, no wiring, and free electricity for its operation (*Siemens Solar*).

Attic vent fans

Lighting and battery charging are not the only photovoltaic applications for the home. Solar-charged attic vent fans are a unique way to save energy. Since heat rises naturally to the highest point in any building, it is estimated that consumers can save 15 percent or more on their electric bill by venting the heat out of the attic. A solar-powered vent fan works the same as does a conventional electric-powered vent fan. It is mounted in the attic or a crawl space near the peak of the house. A hole is cut for the fan casing, and a two-square-foot solar panel is mounted flush on the south-facing roof. When the sun comes up, the fan turns on, and as more sun is available the fan naturally runs faster. The heat vented from the attic during summer months or other periods of warm weather will save energy, help stop the formation of mildew, vent pollen and other air-borne contaminants, and maintain healthy air flow into and out of the house. Most vent fans have switches to allow the homeowner to turn

A new PV-powered attic vent fan was recently displayed at a national builders' show in early 1991 (*Applied Energy Technology Ltd.*).

them off during colder periods, and adjustable slats which close to prevent loss of heat. The vent fan will pay for itself quickly by saving on air-conditioning costs. These fans may also be used in the kitchen to get rid of odors, but in this case they will need a battery to work during the evenings.

Can I produce all the power for my home?

It may surprise you to learn that photovoltaics can even be used to provide all of the power for your residence, summer home and business. In fact, there are thousands of homes in the U.S. that use photovoltaic cells for power. Most of the homes are in remote rural areas located throughout the country. These systems are also widely used in other countries. In Spain, for example, there are more than 4,000 photovoltaic systems on vacation homes.

At the present time, the most cost-effective use of photovoltaics is for a building which is not connected to the electric utility grid. In many of these cases, the solar system cost will equal the amount to be paid to the utility company to connect your house to their electric wires.

A solar-powered house in Southern California (*Photovoltaic Energy Systems, Inc.*).

Remote vacation cabins are ideal sites for PV power systems (*Photovoltaic Energy Systems, Inc.*).

In most states, if you are connected to the power grid, you are allowed to sell the extra electricity that you do not use from your solar system back to the utility through the federal Public Utility Regulatory Policies Act (PURPA). You should contact your state's Public Utility Commission or the state government solar contact listed in chapter 10 for the rules regarding solar utility-interconnection in your state. In some cases, the utility will install a "reverse" electric meter which will credit you for the extra electricity you generate by just running your electric meter backwards. Other utilities may require you to purchase a separate meter to measure the unused output.

In either case, you're a winner. You can win by being a power producer through the generation of electricity to the utility company, or by saving the cost of connecting to the electric grid.

A remote location is an excellent site for PV power (*Siemens Solar*).

How do I choose the right system for my home?

Several companies manufacture special packaged photovoltaic systems for use in cabins and summer homes, as well as regular residences. In each case you will have to determine which is cost-effective and meets your needs. If you have a cabin or weekend home not already connected to the utility company, photovoltaics is most likely to be cost-effective.

However, you will need to determine your demand (need for energy), quality of energy (wattage), and output of energy (voltage). To make your energy go further, special low-voltage lighting and appliances can be used with the system. It has been estimated that there are more than 10,000 homes in the United States totally powered by solar energy, and most of these homeowners report little or no sacrifices in their lifestyles. They have televisions, VCRs, refrigerators, and other appliances, all powered by photovoltaics.

These systems have proven to be highly reliable and run very smoothly. In fact, one of the oldest solar-powered houses in the United States is at the Florida Solar Energy Center. This 5-kilowatt system is on a three-bedroom, two-bath house, and has been in constant use and operating reliably since 1980. The system supplies about 640 kWh of AC power every month.

As with other types of solar systems, it is absolutely essential that you and your contractor size the system properly to meet your needs. Because photovoltaic systems are modular and can be assembled in any size, they must be constructed to meet your needs. This chapter only discusses the prepackaged systems and how best to select them for your optimum use. Keep in mind that your situation may call for a special design or size because of your energy usage. That's easy to do, but it must be measured carefully to give you all the power you will need.

Economics

To calculate the economics of a residential photovoltaics system, you need to be aware of the amount of electricity per day you will need, which is expressed in kilowatt hours per day (kWh/day). Photovoltaics expert Paul Maycock suggests that you can calculate this information from the power needs of the appliances you have in a cabin, or, if your dwelling is connected to the utility, take your monthly utility bill and divide the number of kilowatt hours used by thirty to get the average daily usage. Because electricity usage fluctuates during the year, it would be best to add up a year's worth of electricity used and divide the aggregate kilowatt hours by 365.

Look at the insolation map on page 82 and find your approximate location to determine the amount of sunlight that

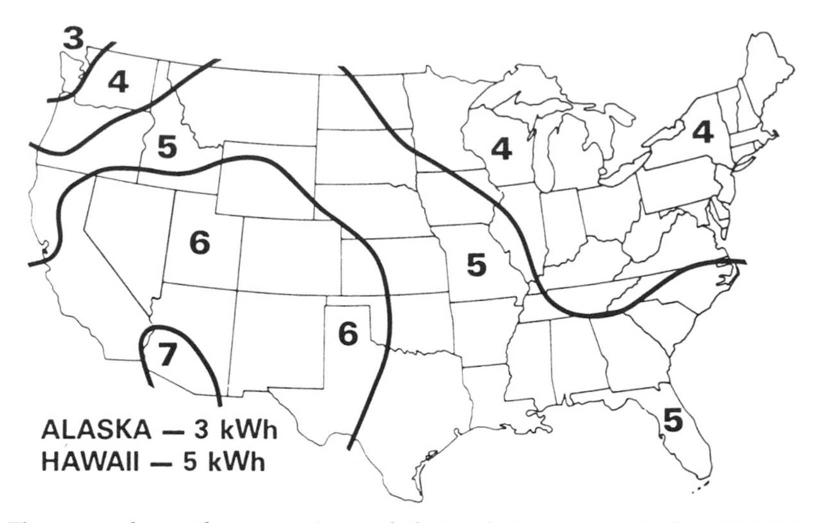

This map shows the approximate daily insolation (amount of sunlight) in kWh/square meter that falls on a south-facing surface for an average day (*Solar Energy Research Institute*).

falls in your area. This map indicates the kilowatts per square meter (approximately per square yard) falling on a south-facing solar panel angled at forty-five degrees. You then need to divide your average daily electric load requirement by the insolation (the amount of sunlight that falls) from the map, and then multiply the result times the system efficiency, which we will assume to be on average 12 percent.

As an example, assume that you have a cabin on a lake in Georgia that you and your family use just about every weekend. The average utility bill shows about 450 kWh of energy usage monthly, or 15 kWh/day. The insolation shown on the map for this area is 5 kWh/square meter/day. Your calculation is a simple one (shown on page 83). To find the required size of the photovoltaic array in square meters, multiply the daily insolation in square meters per day (5) times the PV system efficiency (0.12), and divide this into the number of the average daily load requirement (15 kWh per day). In this example, the result is 25 square meters.

Since one square meter equals 10.76 square feet, you would need 269 square feet of solar electric modules or panels for this system. The average size of a solar module is 12 square feet, so you would need about twenty-two solar modules for this particular home.

To determine the average cost of a solar photovoltaic system, you need to determine the output of the solar system at noon time (per peak kilowatt). If you multiply the system efficiency (again, assume 12 percent) times the total square meters in our example (25), you get 3.0 peak kilowatts. Assume that the installed price (including batteries, solar panels, electric wiring, and controls) is about $4,000 per peak kilowatt, so a 3.0 peak kilowatt average residential photovoltaic system would cost about $12,000.

There is another calculation to estimate the cost of electricity over a system life of twenty years. Multiply the 15 kWh per day times 365 days per year times twenty years (answer is 109,500). Divide this into the $12,000 system cost, and the result is $0.1095 per kWh. This means that the system for this

CALCULATING THE ECONOMICS OF PHOTOVOLTAICS.

Average utility bill

$$\frac{450 \text{ kWh}}{30 \text{ days}} = 15 \text{ kWh/day}$$

Size of PV Array

$$\frac{15 \text{ kWh/day}}{5 \text{ sq. meters/day} \times 0.12 \text{ efficiency}} = 25 \text{ square meters}$$

Output of System

$$0.12 \text{ efficiency} \times 25 \text{ sq. meters} = 3.0 \text{ peak kW}$$

Cost of electricity over system life

$$\frac{\$12{,}000 \text{ System cost}}{15 \text{ kWh} \times 365 \text{ days} \times 20 \text{ years}} = 0.1095 \text{ per kWh}$$

home in Georgia would cost eleven cents per kilowatt hour. The average cost for electricity in the United States ranges from four cents per kWh to twenty cents per kWh, with the nationwide average probably around seven or eight cents per kWh. This means that this solar system is about three cents per kWh higher. However, as stated earlier, the photovoltaic residential system may be able to pay for itself immediately if you have to pay the utility a fee (possibly ranging from $2,500 to $30,000) for running electric wires to your home.

Once a system has been installed, you can greatly reduce costs of energy use by using low-voltage fluorescent lighting and special low-voltage (12 or 24 volt) appliances, including refrigerators, televisions, and blenders which are sized for automobile and trailer use.

Photovoltaics may be very cost-competitive in areas of the country where utilities charge higher electric rates during midday hours. In some areas, these midday rates can top twenty-five cents per kWh! This rate structure would increase the cost-competitiveness of your photovoltaic system dramatically.

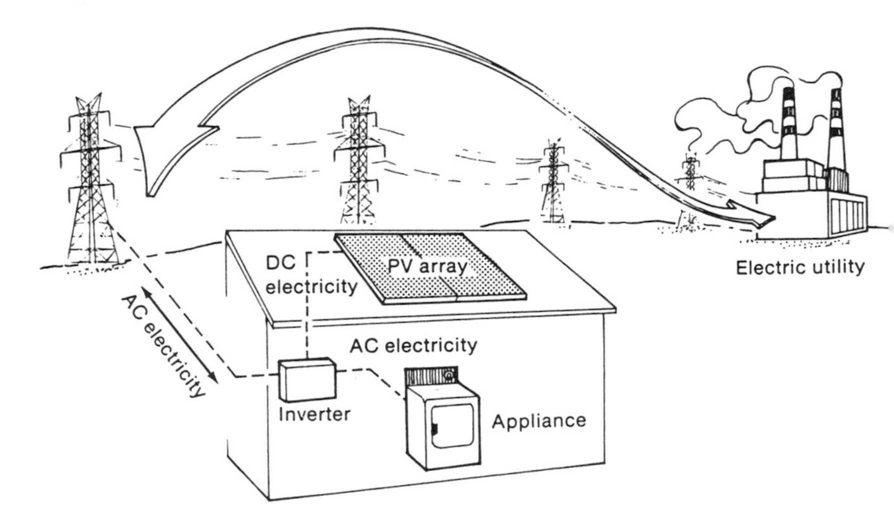

Residential PV systems send excess electricity onto the power grid, but draw from the utility grid when there is not enough sunlight (*Florida Solar Energy Center*).

Another option would be to hook-up a supplemental photovoltaic system to your existing household. In areas with high peak power rates during midday, or in areas where power outages are frequent due to poor utility load management, hurricanes, floods, etc., a photovoltaic system sized to meet the midday peak could be very cost-effective. Some companies market grid-connected photovoltaic systems in the $5,000 range which would provide power to your house. If you need more power, the regular electricity would automatically kick in as a back-up. When your electricity requirement falls below what the photovoltaic system can deliver during the day, the photovoltaic system will take over completely. For instance, if you and your spouse both work and the house sits empty all day, the photovoltaic system will maintain power for your refrigerator and other appliances.

PV is an ideal power source for remote weather monitoring stations to keep people informed of local conditions (*Photovoltaic Energy Systems, Inc.*).

On weekends or other times when the family is home all day and uses more appliances, conventional back-up power can be used. Some of these systems come with battery back-up while others do not.

In either case, you will need to ensure that you have reputable equipment and contractors (see chapter 9 for guidelines on choosing contractors and other professionals). Most people who use photovoltaics for residences do so to save high utility hook-up fees, to assure an electrical supply at all times, or to have a nonpolluting source of energy with no price increases for at least the minimum twenty-year lifetime of their system.

Whatever the reason, photovoltaics do supply a sure and dependable source of electric power. Equipment which meets Underwriters Laboratories (UL) standards is manufactured by over twenty companies in the United States. The equipment has been used in hundreds of commercial and industrial applications in the United States and many residential uses all over the world. The time is coming when you may find photovoltaics the best option for your needs.

CHAPTER SIX

Water Purification

The sun can make water safe to drink

It wasn't too long ago that people took their drinking water for granted. They just went into the kitchen, turned on the tap, and there it was—clean, safe, tasty water to drink. But one look at the shelves of supermarkets and hardware stores today, with the many water purifiers, cartridge systems, bottled water dispensers, and other similar products, and you can see dramatically the growing concern with water quality and the pressing need to do something about it.

The headlines in today's newspapers tell the story behind the concerns. We read about chemical wastes in the ground possibly seeping into underground aquifers. We know of the potential hazards of chlorine, a chemical often added to water treatment systems. We read about the growth demands in the Sun Belt states and other fast-growing parts of the country, putting a burden on all utilities, including the water supply. Other problems, including many natural occurrences, have prompted stories of contamination of drinking water throughout the U.S.

As the concerns have grown, many people have turned to bottled water, electrically powered indoor distillers, and several types of filtration systems to clean their drinking water. While all of the available systems offer many benefits, they have one major problem in common—they're much more expensive than the tap water people can use instead. A utility, for example, might charge its customers seventy-five cents for 1,000 gallons of water, while bottled water can cost seventy-five cents for just one gallon! Even though many people know that tap water might have some problems, they often find it economically hard to justify the expense of bottled or purified water, especially since the taste of the water usually doesn't bother them.

Unfortunately, it's what you can't taste that can cause serious harm. Studies of many water supplies have found dozens of chemicals that can cause potential health problems. Reading a list of contaminants in your water supply would probably scare you into action right away. While most water authorities would say that the concentrations are too low to cause a problem, many people are skeptical and are looking for alternatives.

One excellent alternative is a solar distillation system.

It's possible to easily build or fairly inexpensively purchase a solar still which will provide distilled water for your needs. The biggest advantage of a solar distiller is that the sunlight to heat water for distillation has absolutely zero fuel costs. While these systems do require some equipment, they will pay for themselves fairly quickly and provide pure, safe water for many years.

Solar stills are highly efficient in bringing water to the proper temperature. Water doesn't have to be boiled to be distilled. Simply elevating the temperature to a point near the

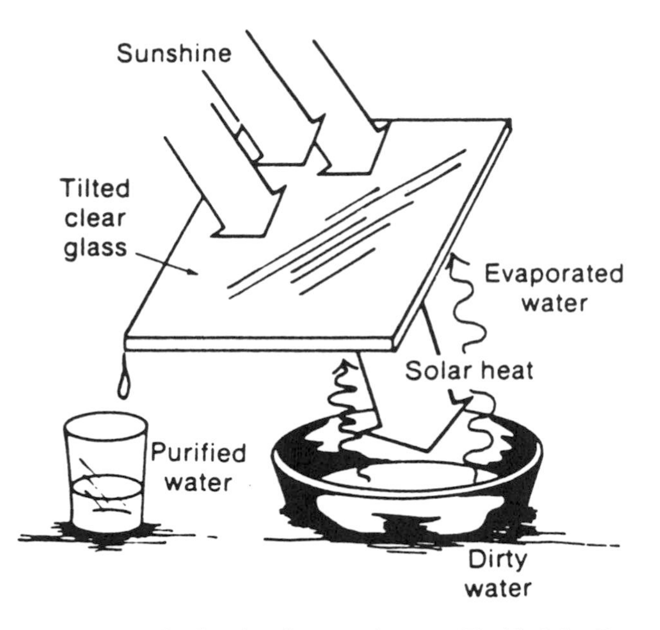

The basic concept of solar distillation of water (*Florida Solar Energy Center*).

boiling level will adequately increase the evaporation rate. Keeping the water below boiling will also ensure that some of the unwanted residue that can get into distilled water through vigorous boiling will stay away from pure water.

The average family can get enough drinking water for their needs from a solar still. In many parts of the world, in fact, large solar distilling units are used to provide water for entire neighborhoods or villages. In towns on islands throughout the Bahamas and in other tropical and subtropical regions, solar stills purify sea water for drinking and cooking. For example, a solar-powered desalination plant in Saudi Arabia produces enough drinking water to meet the daily needs of a community of 250 people. There are records of simple solar distillation stills being used since the nineteenth century. Historians even tell of a solar still built almost 100 years ago in the high country of Chile for mine workers in an area without pure drinking water. That homemade system produced as much as 6,000 gallons of fresh water every day.

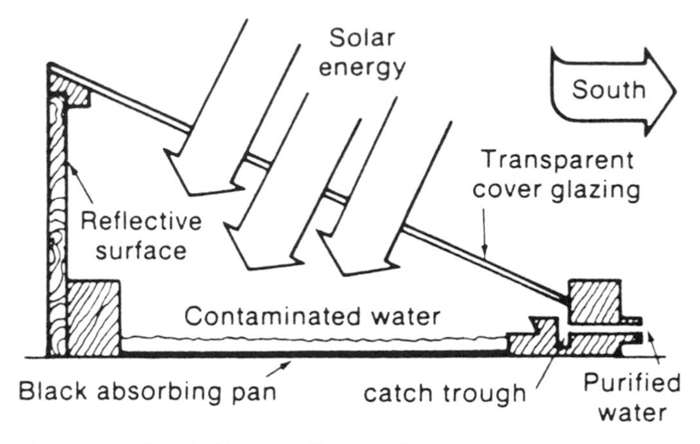

A simple solar still (*Florida Solar Energy Center*).

A basic solar still is a very simple device. If salty or contaminated water is left in an open container, the fresh water will evaporate into the air, separating itself from the salt, dirt, algae, or other organic pollutants in the water. A solar still uses solar energy to speed up the evaporation rate and captures the evaporated water by condensing it onto a cool surface. A solar still, then, is essentially an insulated, dark-colored container or shallow pan, covered by a sheet of clear glass or plastic that is tilted slightly to let the fresh water that condenses on it trickle down into a collection trough. The still is filled with six to twelve inches of water, and the water is evaporated by solar thermal energy, which condenses the water vapor onto the glazing material. The glazing also serves to hold the heat inside of the unit.

Several companies in the U.S. make simple solar distillers that can provide about two gallons of distilled water per day in sunny weather. These low-maintenance, fuel-free units remove just about every type of contaminant from the water, including chlorine odor and taste, bacteria, sediment, and sodium.

One manufacturer even makes a solar water heating unit that produces drinkable water as a by-product. In areas like Southern California, for example, it has been estimated that as many as one-third of the people buy bottled water every week. The new solar water heater now on the market produces about two-and-one-half gallons of distilled water every day, along with enough hot water for the average family, and is priced comparably to a conventional solar water heating system.

The slow solar distillation process is very thorough, allowing only the pure water to evaporate and collect in the trough. At the present time, the economics of solar-distilled water do not make it competitive with utility-supplied drinking water. However, it is highly competitive with bottled water. The average family using bottled water regularly pays $2 to $4 a week for bottled water delivered to their home, or up to $210 per year.

Finally, campers and owners of summer homes and other rural properties should know that photovoltaics can be used for water disinfection. Solar systems are used to treat rural water supplies in many developing countries because the systems are simple and require little maintenance. This same technology can be used by consumers in the United States.

Photovoltaic water disinfection and purification systems are best used for nonelectrified areas. Many wells and streams are contaminated, and you should have them tested if your property has one. If there is a problem with the water, a PV water system may be the answer you need.

A conventional method for disinfecting water is to add chemicals or use a disinfectant-generating device. However, the supply of chemicals may not be reliable and can be difficult to handle and store. One alternative is an electrolysis cell, a device containing salt water that is electrolyzed to produce chlorine and oxygen oxidants to disinfect the water. A PV power supply can easily be used for this type of system. Because the oxygen oxidants react quickly, there is usually no bad taste or odor from the chlorine.

There are also several types of mixed-oxidant generators available for use today which are more effective against such organisms as those causing Legionella Pneumophila and Giardia cysts and other hard-to-kill bacteria.

Small solar water distillers like this unit can provide two gallons of pure water daily (*Solar Development Inc.*).

Photovoltaic-powered water purification and disinfection systems can meet the needs of many consumers in areas around the country. If you spend a lot of time camping or travelling in very remote areas, you may wish to consider using a simple PV system to protect your health and safety.

As the cost of pure water increases and more people worry about the quality of their tap water, solar stills will play a more important role in meeting one of life's basic needs—safe, clean drinking water.

CHAPTER SEVEN

Solar Cooking

Using the sun to cook your meals

Some of the most energy-intensive appliances in your home are located in your kitchen. Energy-efficient refrigerators, freezers, and ovens have been introduced in recent years, but even these still consume a great deal of energy. Now advances in solar technology make it possible to use solar cookers for some of your needs. While these appliances are most suited for remote applications—at campgrounds and weekend cabins, on a boat or at the beach, and other places away from home—many people use them for weekend barbecues and, on occasion, to supplement their conventional appliances.

How often during the summer do you use an outdoor barbecue grill? Have you ever taken a camping trip and used a portable charcoal cooker? Now you can replace these items with solar-powered appliances and end your reliance on fossil fuels—fuels which are often unavailable at any cost when you're away from home.

Standing in front of a hot oven, especially in the summertime, is one of the most unpleasant household chores. It's bad enough having to cook when the temperature climbs to ninety degrees outside, feeling all that heat from the oven and growing more uncomfortable each minute. But what's even worse is the realization that all that hot air must be cooled off, working your air conditioner even harder while you cook. One estimate says that for each dollar spent on conventional cooking in an air-conditioned home, three more dollars must be spent to cool the house back down. But that hot sun that's costing you so much money in air conditioning could save you some money by doing the cooking for you.

It's been a little more than 200 years since a Swiss scientist named Horace de Saussure invented the first solar cooker. While it's logical to assume that people have used solar power to start fires and cook meats for thousands of years, de Saussure's research into solar radiation probably started the commercial solar cooking industry.

While studying the effects of altitude on solar radiation inside insulated boxes covered with glass, the scientist noted that the high temperatures in the boxes could be used for practical purposes such as baking. He tried cooking some fruits inside the boxes and found that he liked their taste.

Now, many years later, a number of manufacturers have designed variations of the "hot boxes" which are being used worldwide for cooking and baking. In many developing countries, especially in places where they have deforested most of their region, these solar ovens are the only way people can cook meals, boil water, and otherwise heat foods.

You can take advantage of the same technology being used in many parts of the world to help with your cooking needs all year long. Many solar chefs have commented on their preference for meals cooked in solar ovens rather than

Reflectors on solar ovens concentrate sunlight for higher temperatures (*Florida Solar Energy Center Library*).

in conventional ovens. They enjoy the taste of food that is gently cooked by the sun rather than bombarded with hot air in a conventional oven. Solar ovens cook cleanly and without burning, and allow the food to retain most of its vitamins. And cooking takes about the same length of time as when using conventional ovens.

Solar ovens work on the same principle as that of solar collectors in domestic water heating systems. Both systems collect solar radiation, convert it to heat, and then transfer it to another substance. The ovens, however, absorb the sun's rays and use reflective surfaces so the sun's energy is aimed at the food.

There are three basic types of solar ovens—box, parabolic reflector, and multi-reflector.

Box ovens are the least expensive, but they produce the lowest temperatures, usually getting no higher than 275 to 300 degrees Fahrenheit. These are simply well-insulated boxes that allow sunlight to enter through a glass-covered opening. The slow cooking makes this oven suitable for slow boiling of foods such as stews and cereals. It can also be used to warm sandwiches and cook foods like hot dogs and pot pies. Turning the box every hour to orient it toward the sun will improve its performance. This is the most popular type of solar cooker because of its simplicity, low cost, and ease of transportability. Most units are large enough to hold two or three cooking pots and cook ten to fifteen pounds of food on a sunny day. A few manufacturers produce these units commercially.

Parabolic reflector ovens produce higher temperatures, but must be constantly readjusted to focus directly on the sun. These ovens work by focussing the sun's rays onto the cooking area. Only kits for building these are available at the present time.

Multi-reflector ovens are generally considered the best overall because of their capacity and temperature range. They can achieve higher temperatures and be used for a wider variety of foods, including meats where high temperatures are needed.

Solar ovens can bake breads, cook meats, and heat sandwiches. Most ovens fold up for easy portability (*Burns-Milwaukee, Inc.*).

Solar ovens and cookers are still basically novelty items in the United States because we have other fuel sources for most of our needs, and the energy costs are not terribly high. However, they are still functional and efficient, and attractive to homeowners for a number of reasons. Cooking outdoors keeps the kitchen cooler. The ovens will be more economically attractive as energy costs go up. And most of the cookers are fairly portable so you can take them camping, to the beach, or wherever you want to go. Solar cookers are especially desirable to campers and hikers, since the food is covered while cooking—safe from dirt and hungry animals. You can also leave the oven alone while the meal is cooking and come back to it when the food is ready. And many campgrounds and natural parks ban fires, making solar cookers the only alternative for hot food preparation.

There are hundreds of success stories of people in remote locations using solar cookers to prepare their meals. Peace Corps workers and other volunteers in more than 100 countries around the world use solar ovens for cooking and baking, as well as to sterilize medical instruments and purify water. And consider the 1990 American Everest-Lhotse Expedition, whose members used a solar cooker at their base camp to bake bread—at an altitude of 17,600 feet!

Twice a month on Friday mornings, dozens of people stop by the East Yolo Senior Club in West Sacramento, California, and enjoy a "solar potluck" meal. For the past four years, these get-togethers have been held on a regular basis for the hundreds of people in the area who have built their own solar ovens. When the program started, the club held its first solar cookout which featured thirty-five solar ovens built by members of the club.

A few years before the club started their potluck meals, the River City High School Homemaking Club held a Solar Foods Festival at a local church. In fact, it has been estimated that 1 percent of the homes in West Sacramento use solar box cookers. This is an area where the weather affords about 200 cooking days every year.

Though many people make their own solar cookers because the basic design of a unit is fairly simple, most people who use them often buy them rather than make their own. There are a few manufacturers, mostly located in the western United States. Prices range from around $50 to $300 or more, depending on the type of cooker and its features. Contact some of the organizations listed in chapter 10 for the names of manufacturers and distributors.

Solar cookers aren't perfect for all jobs, though. For example, most home economists recommend that you don't cook fresh pork, since the cookers won't produce high enough constant temperatures over long enough periods of time to kill trichinosis. Another limitation concerns the eye appeal of some foods. Ultraviolet rays will fade green, orange, or yellow vegetables to a brownish or white color, and take away some of their vitamin content as well.

A young girl in La Baritta, Mexico, cooks the evening meal in a solar oven (*Burns-Milwaukee, Inc.*).

However, if the sun is bright enough, cooking times in solar cookers can be comparable to conventional ovens. On cloudy days, it is important to use containers that hold the heat well so that you can cook even without bright sunlight. Depending on the design of the cooker, you can get temperatures of up to 500 degrees. Adjust the cooker toward the sun and the temperature will be maintained at the desired setting. These cookers work well in winter, too, since it is the amount of sunshine, not the temperature of the air, that cooks the food.

One of the best uses of solar cookers is for baking, since most baked goods need only moderate temperatures. Cakes and chewy dessert recipes such as brownies come out very well in solar cookers. There are other benefits as well. A college professor conducting a study on the use of solar cookers in Zambia wrote a research report showing that solar-cooked foods were nutritionally as good or better than the same foods cooked by traditional methods. People on special diets find solar-cooked food desirable because it is best cooked in its own skins (e.g., corn in husks or potatoes in skins). And many homeowners report that everything from barbecue ribs to pizza tastes better when cooked by the sun.

To get you started, here are a couple of easy solar recipes from Jo Townsend of the University of Florida Extension Service. Write to some of the organizations in chapter 10 for more ideas. Keep in mind that any conventional recipe is suitable for a solar oven—just adjust the time, if necessary, to take into account the lower temperature.

Lemon Chicknic
2 T butter or margarine, melted
1 clove garlic, crushed
4 chicken breast halves, skinned and boned
1/4 cup Italian-style dry breadcrumbs
3 T freshly squeezed lemon juice
Lemon slices
Fresh parsley sprigs

Combine butter and garlic in a baking dish. Place in the oven to melt butter. Flatten chicken with a meat mallet; fold envelope fashion and roll in breadcrumbs. Place chicken in baking dish, turning to coat well. Slide dish into a browning bag so steam can escape. Bake forty minutes or until chicken is tender. Pour lemon juice over chicken and bake an additional twenty minutes, basting occasionally with butter and lemon juice from dish. Garnish with lemon and parsley. Makes four servings.

Backyard Baked Beans
2 slices bacon
16 oz. can (1 3/4 cups) pork & beans
1/4 cup firmly packed brown sugar
1 small chopped onion
1 t prepared mustard
1/4 cup catsup
2 T Worcestershire sauce

Cut bacon into small pieces. Combine chopped onion and bacon in iron dutch oven, cover with lid. Cook until bacon is brown and onion tender. Combine remaining ingredients in the dutch oven. Bake covered for one hour or until beans are thickened and heated through. Leave corner of lid cracked so steam can escape. Lid prevents splatters. Makes four servings.

Solar S'Mores
1/2 cup crunchy peanut butter
12 graham crackers, halved
6 large marshmallows

Spread peanut butter on 6 graham crackers, top with marshmallows and place on oven rack. Cook for ten minutes or until marshmallows begin to melt. Cover with remaining graham cracker squares to form a sandwich. Press to seal. Serve immediately. Makes six servings.

To learn more about solar cookers, write to Solar Box Cookers Northwest (SBCN) and ask for a copy of their quarterly newsletter, *Solar Box Journal.* A recent issue included news briefs on the use of solar cookers around the world, plans for building new types of cookers, and recipes for solar cookers ranging from herbal vinegar and songbird cornbread to herbal skewers and even solar dog biscuits. Write SBCN, 7036 Eighteenth Avenue, N.E., Seattle, WA 98115. Also write to the national solar associations for listings of manufacturers and distributors. You can be enjoying a "solarque" on picnics this summer!

CHAPTER EIGHT

Solar Cooling and Space Heating

Keeping your home comfortable all year long

Most uses of solar energy involve mechanical equipment that captures sunlight and uses it to produce heat or electricity. This requires the purchase of specialized equipment, installation of the equipment, and maintenance and repair as needed.

There is one important use of solar energy, though, that doesn't call for any special equipment at all. The term "passive" is used to describe how solar building designs and materials can provide cooling and heating to keep a home comfortable and energy-efficient without the use of mechanical equipment. This style of construction results in homes that respond to the environment.

In the case of passive cooling, the

A passive solar home in New Mexico (*Solar Energy Research Institute*).

plan of a house, careful site selection and planning, construction materials, building features, and other aspects of the home are designed to block the sun's rays in summer to keep the home cooler. In passive heating, these same basic features are designed to collect, store and distribute the sun's heat in winter inside the home to improve wintertime comfort. When you consider that nearly half of the residential energy use in this country is for space cooling and heating, you realize that this represents an area with significant potential for saving money on your utility bill.

The key principle behind passive design is solar motion—the movement of the sun through the sky during the year. Solar motion has been studied and revered by man since his arrival on earth, as is evidenced by Stonehenge in England. Three factors—time of day, date of the year, and latitude—combine to place the sun at a different position in the sky at every hour of the half year. With the exception of the summer and winter solstices, on any given day of the year there is one other day of the year when the sun's path will be identical. In addition, the sun will be in the same relative position in the western hemisphere of the sky in the afternoon as it was in the eastern hemisphere in the morning.

Further, the sun rarely rises due east and rarely sets due west. It usually is either to the north or south of these directions when it rises and sets. The solar motion phenomenon provides a natural framework for the design of good buildings which can protect the south side in summer and allow exposure of that side in the winter.

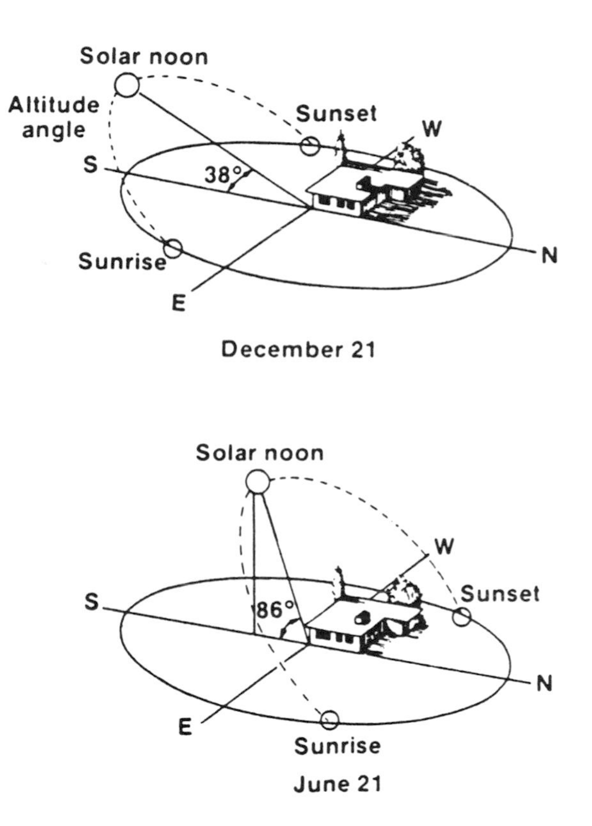

In winter, the sun is lower and more southerly than it is in summer (*Florida Solar Energy Center*).

More than 2,500 years ago in ancient Greece, entire cities were built to take advantage of the sun and the climate. Roman ruins show evidence of villas and bath houses that were heated by solar energy and used large expanses of daylighting and ventilation throughout the buildings. Drawings and photographs of the cliff dwellings of early American Indians, the sod homes of the pioneers, and Eskimo igloos show the uses of building homes to respond to the environment. Even Thomas Jefferson's home in Monticello took into consideration solar and wind effects to increase comfort. In many cases, this type of design was a necessity to allow the occupants to live comfortably and safely within the confines of temperature extremes and harsh weather.

Until the development of efficient heating and cooling systems and the widespread availability of inexpensive fossil fuels after World War II, homes in the U.S. were usually built to take advantage of passive solar principles. However, once fuel was easily affordable and available, many homebuilders turned away from these principles and relied more extensively on mechanical cooling and heating systems. The energy crises of the 1970s and 1990 again focused national attention on the use of passively designed homes to reduce energy usage. Research has shown, in fact, that about 35 percent of the total energy used in the United States is consumed by buildings. This is an area where significant savings can be achieved.

During the past twenty years, there has been extensive research on energy use in buildings, resulting in a great deal of new information for architects, builders, and developers. While passive solar heating has long been an area of interest, much of this new information looks at passive cooling and offers energy savings for the rapidly growing population in the Sun Belt states. However, these principles can be used almost everywhere in the United States to achieve savings in the hot weather. Generally, areas in the south will save more by using passive cooling strategies, while northern climates offer the greatest potential savings from passive heating strategies. Both areas can still take advantage of the other's principles, however.

The demand for energy-efficient buildings is on the increase in the United States, and energy-conscious design has become a major part of today's building profession. The term "affordable comfort" is often used today to refer to passive design principles that minimize a home's energy consumption while maximizing its livability level. Homeowners in all parts of the country can take advantage of passive solar building strategies to significantly lower their utility bills while making their homes comfortable all year long.

It has been estimated that there are now more than a quarter of a million homes in the United States that have been built with passive solar features. In most cases, these are new homes that have been designed and built with passive strategies incorporated into the basic plans and design. However, in many other cases, homeowners have been able to remodel their existing homes to take advantage of passive strategies. Research has shown that these strategies can help homeowners reduce their utility bills by 10 to 40 percent with only minimal increase in renovation costs.

One of the best benefits of passive solar homes is that they are much more comfortable. By properly locating a home on a lot to take advantage of the sun's warmth, homes can be planned with better views and more enjoyable use of the outdoor areas. Homes that are well-insulated and tightly sealed have quieter interiors, fewer drafty areas and "hot spots," and more time during the year to take advantage of open windows, fresh air, and more use of natural ventilation.

Depending on the climate where you live, you can choose the best passive cooling or heating features for your home. Note that some of these will work only in new construction, and are difficult or even impossible to add after the home has been built. Other features, though, can be added in a "retrofit" situation to make an existing home more comfortable and energy-efficient.

Finally, a general passive strategy called daylighting can be used in homes to improve comfort and reduce energy use all year long. First, this strategy of using natural light in a building will cut down on energy use for lighting. Large expanses of windows, clerestory windows, skylights, and other clear areas allow the sun's warmth into the house during the winter to provide more heat for the house. Blocking these areas with shades, drapes, blinds, and other forms of movable insulation in summer will reduce heat buildup and control the amount of light entering the home. Second, daylighting saves energy used for cooling to lower the heat buildup caused by incandescent light bulbs. Because as much as 95 percent of the energy given off by a light bulb is in the form of heat rather than light, the cooling system has to work to reduce this unwanted heat.

A passively cooled home (*U.S. Department of Energy*).

Passive cooling

Whether you live in the dry areas of the desert Southwest or the hot, humid Southeast, or anywhere from the muggy mid-Atlantic states to the nation's midsection, passive cooling strategies can keep your home more comfortable in the hottest summer months while reducing your dependence on air conditioning. The following strategies offer the best potential for maximum savings and comfort.

Ventilation

The key element in passive cooling is ventilation. This includes designing your home to allow available breezes to flow through the house, building in attic vents to allow hot air to escape before it overheats your home, and installing movable windows to control airflow as weather conditions change.

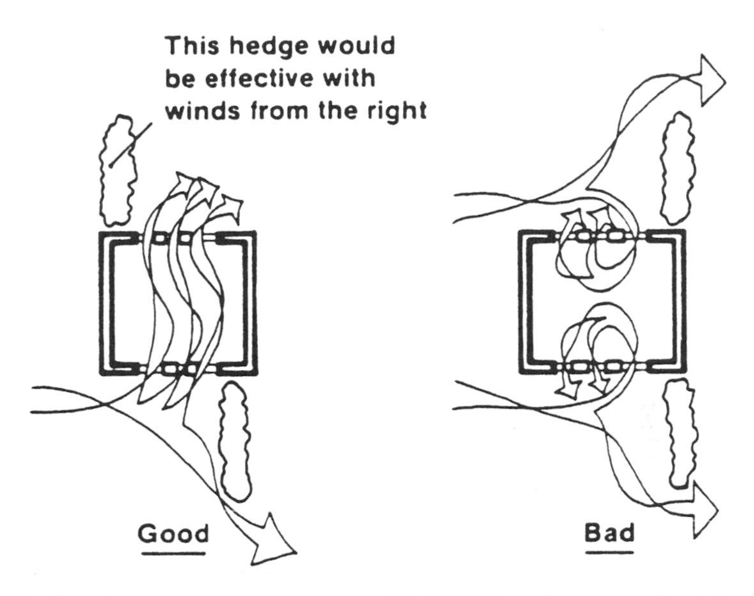

Windbreaks can promote cross-ventilation. The design on the left works well; the one on the right will not cross-ventilate (*Florida Solar Energy Center*).

Trees can be planted to guide winds through and around the home. Rooms and partitions can be built to create a "wingwall" effect that pulls air into the home and helps circulate it through other outlets. Rooms can be designed and planned to maximize the use of breezes and wind currents.

Fans are another very effective way of keeping your home comfortable while maximizing natural breezes. They are especially useful for keeping people comfortable in hot climates in the "swing seasons" just before and after the peak months when air conditioning is most used. Fans circulate air on and around the people in a room to keep them more comfortable, and they do this while using less energy than does a typical light bulb.

You can also use fans along with your air conditioner to help lower energy usage in the home. Researchers at the Florida Solar Energy Center have estimated that the use of a fan allows homeowners to set their air-conditioner thermostat two to six degrees higher than they would otherwise. Since each degree of increase saves about 8 percent in cooling costs, fans help you dramatically lower your cooling bills during hot weather. In the pleasant fall and spring days, fans let you turn the air conditioner off totally.

Ceiling paddle fans are widely used to increase ventilation in homes (*Florida Solar Energy Center Library*).

During winter, when you heat your house, the heat rises to the top of your room. Using ceiling fans to bring the heat down and into the room will save money and increase comfort.

Options for fans include air circulating fans to create air motion in the home, whole house fans to keep fresh air moving throughout the building, and attic vent fans to lower air temperatures and help keep the house more comfortable.

Site Planning and Home Design

If you're planning a new home, positioning it with room for breezes to flow around it will help keep the indoors cool.

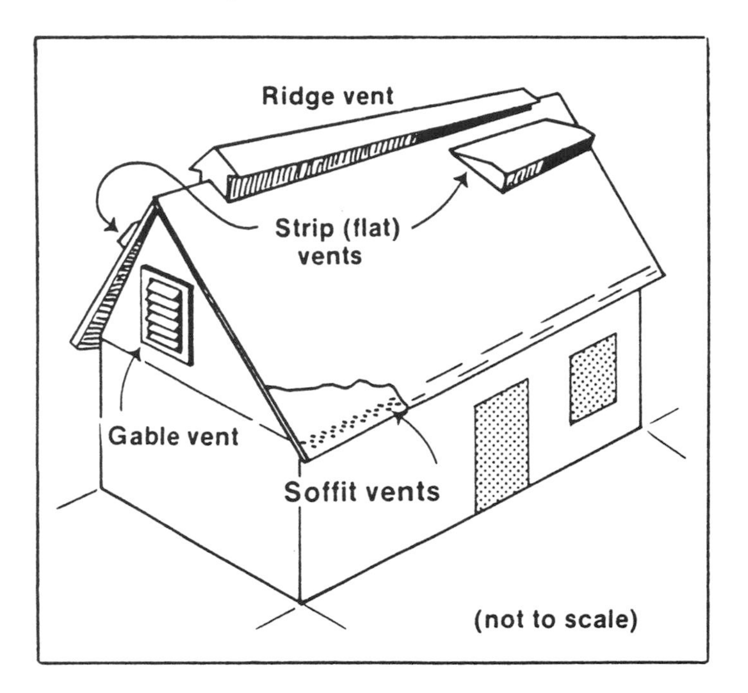

Ridge, gable, and strip vents can greatly increase attic ventilation (*Florida Solar Energy Center*).

Designing the house so the smallest wall area faces the east and west sides will minimize exposure to the hot morning and afternoon sun. Taking advantage of overhangs, eaves, and other shading structures will help block the higher summer sun while allowing the winter sun to provide warmth in cooler weather.

When planning the site and orientation for your next home, consider the following recommendations:

- Have the long axis of the home facing north and south. The size of the lot itself will have a major bearing on the ability of the builder to align your home in an east/west axis, but most sites can accommodate a home in this fashion if planned in advance. Side entry garages can also help relieve the impression of the garage dominating the street side elevation of the home.

- Minimize windows on the east and west sides. The best plan for a house on an east and west facing lot involves having just one front window and back and front porches, with a garage or carport on the side of the home facing north or south.
- Don't have too many windows. Since windows are a significant source of heat gain in summer and heat loss in winter, proper orientation and selection of glass can help improve indoor comfort. As a general rule, builders of energy-efficient homes plan on 15 percent of the floor area for well-spaced windows, which is adequate to provide a well-lit interior. If you want more window area than that, be sure the glass is well-shaded, or consider some of the more elaborate window shading products, insulated windows, selective surface windows, and window films. Many homes in the southern U.S. are quite attractive with 10 percent or less of the floor area as window glass.
- Use porches to shade windows on the east and west sides. Permanent shading with porches or other uses of the building's shape will cut down on the summer heat gain through these windows. Buffer spaces such as closets, laundry rooms, or other rooms where people spend less time can be located on the east and west sides to help minimize excessive heat from the sun.
- Choose plans that provide overhangs on all sides of the house. At the minimum, be sure to have an overhang on the south side to shade direct sun in summer.
- Finally, consider time of day or room usage in finalizing the layout. Many people in southern climates have learned the hard way that west-facing kitchens, dining rooms and porches can be very uncomfortable during the preparation and consumption of the evening meal. On the other hand, locating the living room on the south side of the home allows you to take advantage of winter warmth.

Shading

The sun can do a lot of damage in your home if you don't control it. It can raise the inside temperature, create blinding glare, and fade fabrics and materials. Just closing your drapes or blinds won't do the best job because the sun's energy has already gotten into your house. The space between the shades and the windows will capture much of the sun's heat, eventually distributing that heat into your home's living area.

Both interior and exterior shades, awnings, and blinds can be used and controlled by the occupants to reduce the sun's effects and keep the home more comfortable. Good shading strategies can save you 10 to 20 percent on your cooling and heating bills, but the level of comfort they can bring far surpasses the economic benefits.

Though experts usually recommend outdoor shading because it stops the heat before it gets to the home, they usually add that if outdoor shading can't be used, interior shading is

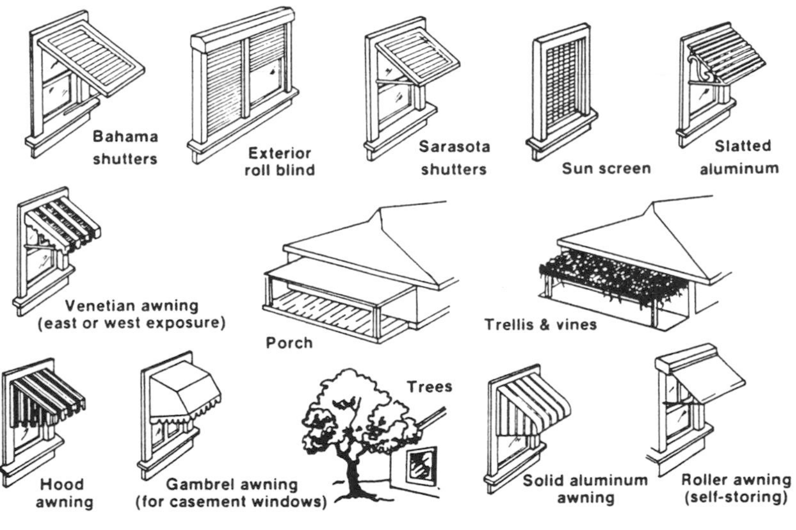

Typical exterior shading options (*Florida Solar Energy Center*).

a very effective alternative. It's better than no shade at all, and in some cases can do an even better job because you're more likely to adjust operable shades that are located inside your home.

The best of all possible shades would be a tall, full tree right outside your window, letting the foliage block the sun's rays. A good second choice for shading would be a tall fence, but in reality it would be tough to make the fence tall enough to block the sun completely. The best alternative is the use of exterior awnings, which will stop much of the heat before it reaches the house but will still allow indirect and diffuse light for illumination indoors. For example, window awnings are available that can be set for maximum shading in summer and then be manually retracted to increase sunlight in winter.

Bahama shutters are widely used in the south to block the sun while allowing ventilation (*Florida Solar Energy Center Library*).

Before making the final decision on window shading options, you need to consider cost, shading efficiency, durability, and aesthetics. Instead of canvas or metal awnings, you might prefer louvered insect screens, window films, roll blinds, or some type of window shutters. There are a variety of models of all types of shades—some adjustable, some insulated, and others fixed in position.

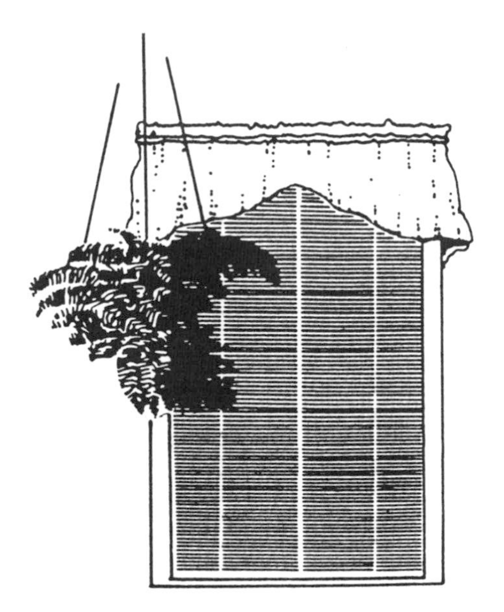

Blinds, white-backed draperies and shades are effective interior shading choices (*Florida Solar Energy Center*).

Because of cost, aesthetics, or property restrictions, you may find outdoor shades inappropriate for your situation. You still need to do something to block direct sunlight, minimize heat gain, and reduce fading and deterioration of fabrics and furniture. The best thing you can do is install tight-fitting shades on tracks mounted along the windows in your home, especially on those facing east or west. The tracks will keep the shades fitted close to the wall and prevent heat from entering the living space. Since the shades are inside, you can raise or lower them as needed. Unfortunately, these shades are among the highest priced of the indoor shading options, so they are out of the price range of many homeowners.

A second choice is to use miniblinds or verticals, which can also be easily adjusted to block the harsh sunlight but also opened when you want to take advantage of the sun's rays. Light colors are usually recommended since they won't absorb heat as much as will dark colors. Aesthetics, though, often win out in the choice of blinds and you may find that darker colors match your room better.

Radiant Barriers

Attic radiant barriers—layers of aluminum foil placed in an attic airspace to block radiant heat transfer—have become very popular in hot climates during the past few years, and offer great energy-savings potential. These products, which are often sold in hardware stores and flea markets, as well as traditional solar energy product outlets, can save you money, keep your home comfortable, and pay for themselves in energy savings in a relatively short time period.

A radiant barrier is actually just a new way of using an old building concept. For many years, home builders in the southern U.S. have used foil materials in attics and walls of homes. Researchers at the Florida Solar Energy Center found that using a foil material—not too different from the aluminum foil you use in your kitchen to wrap food—and install-

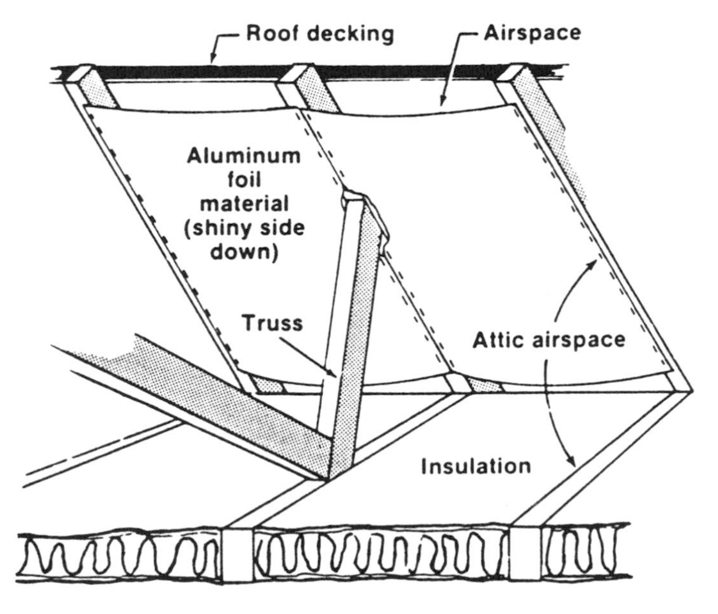

An installed radiant barrier (*Florida Solar Energy Center*).

ing it under the roof rafters so that it faces an airspace, can cut down on about 95 percent of the solar radiation that would otherwise get through to the attic insulation. If not kept away, much of the heat from the hotter insulation would be conducted down into the home's living space.

In a sense, a radiant barrier acts like an umbrella over the insulation, so the product will definitely help lower utility bills. You will save somewhat on winter heating bills by using these products, but the greatest savings come in the summer. The actual energy savings are figured by the amount of load the attic places on the air conditioner. In the Southeast, for example, the attic of a typical home contributes between 10 and 20 percent of the total load on the air conditioner. Studies have shown that heat movement through the ceiling is usually cut by about 45 percent, so homeowners can expect to save between 8 and 12 percent on their cooling bills. Based upon an average selling price of around twenty-five or thirty cents per square foot, installed, this usually translates into a payback of investment in three or four years. Many homeowners find it relatively easy to install a radiant barrier themselves, saving even more money on installation costs. The

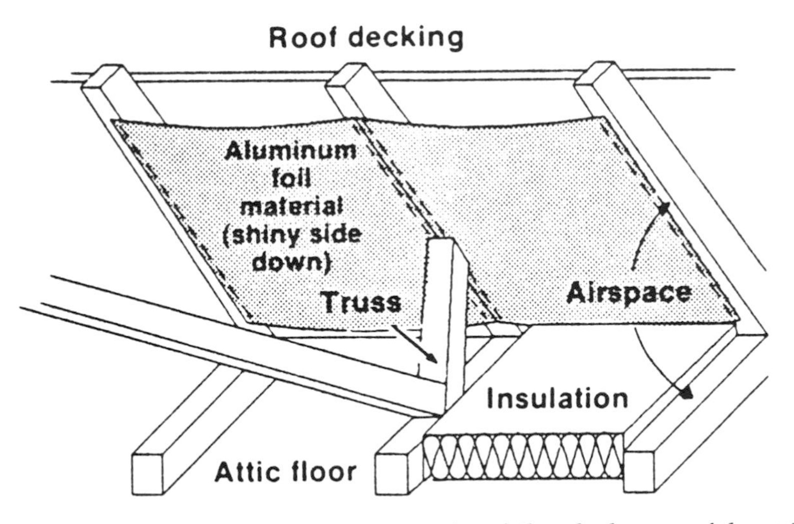

A radiant barrier can be installed by stapling foil to the bottom of the attic trusses (*Florida Solar Energy Center*).

Do-it-yourselfers can usually install a radiant barrier in just a few hours (*Florida Solar Energy Center*).

basic material can be bought for as little as ten to twenty cents per square foot in many locations.

However, the biggest advantage of a radiant barrier may be in the comfort level in the home. In the summer, a sizable portion of the downward heat gain from the attic is blocked, while during milder seasons occupants of the home can stay comfortable without having to use air conditioning. Radiant barriers also expand the home's use of space, as uninsulated rooms like garages, porches, and workrooms become more comfortable when radiant barriers are used in the attic above them.

There is one caution, though. These products have become extremely popular in the South, where they offer the greatest potential savings on utility bills, and many dealers have promised very high savings and unreasonable paybacks. First, keep in mind that a radiant barrier cannot save more than the maximum load. This means that if a roof only contributes to 20 percent of your cooling or heating bill, then the radiant barrier cannot save you any more than this 20 percent figure. Second, organizations do not presently test or certify radiant barrier material. Most foil comes with a number stating its emissivity—how well the product blocks sunlight. But this is usually a number assigned by the manufacturer, not by an independent research lab, so use it cautiously. Finally, researchers usually recommend that you buy the least expensive foil you can find. A single-sided material will work about as well as the fancy and much more expensive double-sided material and the other more exotic foils sold by some companies. Think about it this way—if the basic foil will stop 95 percent of the heat transfer, other materials can only stop a percentage of the remaining 5 percent. This usually makes it uneconomical to buy anything more than you really need.

One other product on the market that you might consider for this purpose is a spray radiant barrier paint which can be used instead of the foil material. The silver-colored paint can be sprayed in the attic under the roof decking with a pressure gun. Unlike standard aluminum paint, this material is made of low-emissivity material that helps reflect the

A spray radiant barrier paint is an alternative to the foil systems. Here the paint is applied to the rafters during home construction (*Solar Energy Corporation*).

sun from the roof decking. Check with your local hardware store or energy company for more information.

Landscaping

Take a look at housing developments in your city. Do builders typically cut down all of the trees and create subdivisions without trees and shrubbery? This has happened in many parts of the U.S. As a result, homeowners must plant new landscaping and shade trees, which can take many years to reach sizes that help block the sun. If you're planning a new home, talk with your builder about saving the trees and shrubs on the lot. It makes a lot more sense and pays bigger dividends if you can leave the natural landscaping in place.

Trees can be very beneficial in shading roofs and walls, directing breezes through the home, stopping reflected heat from nearby gravel and other radiating surfaces, helping keep the air supply clean by trapping and filtering dust and removing carbon dioxide, and helping lower air temperatures around the house. This cooler air can be drawn into the

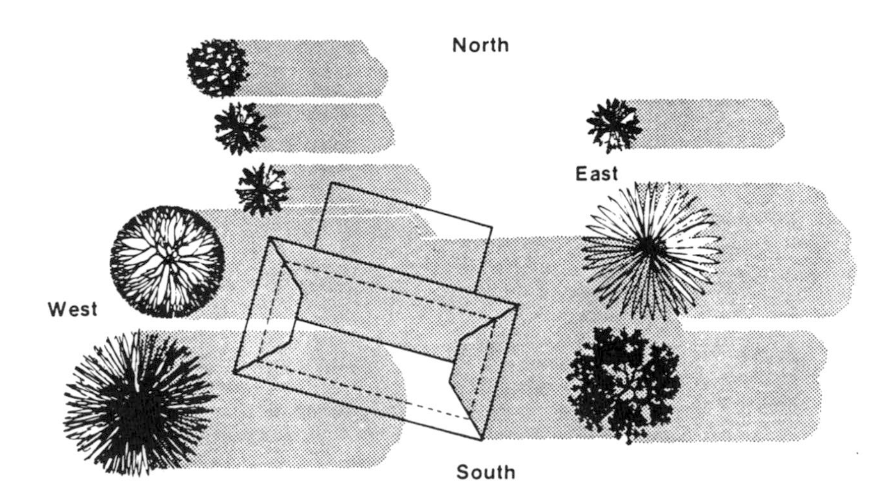

Trees can help reduce solar heat on walls and windows early in the morning and late in the afternoon (*Florida Solar Energy Center*).

house and provide natural cooling. The landscaping thus produces warm, sunny winter spaces and cool, shady summer spaces.

Plants also help produce a natural cooling effect through evapo-transpiration, which is the process whereby a plant cools and humidifies the air adjacent to it. A full tree canopy gives a much cooler ground environment, improving the sensation of comfort in the home. One study found that a single mature tree can get rid of as much heat as would five 10,000 Btu air conditioners. On hot summer days in the South, walls shaded by trees have been found to be as much as twenty-five degrees Fahrenheit cooler than nearby unshaded walls.

Dense landscaping can be effective in channelling summer breezes to improve building ventilation while deflecting harsh winter winds to provide warmth. One passive cooling strategy calls for planting a staggered row of dense evergreens north of the home to help block winter winds, and deciduous trees along the southeast side to block summer sun when the leaves are full but to allow the warming winter sun when the leaves fall off. Climbing vines on exposed east- and west-facing walls absorb and reflect solar energy in summer, and provide immediate shade while you wait for newly planted trees to mature to suitable heights.

To effectively use landscaping to help cool your home, your first priority should be to block any windows on the east and west walls, since this will produce the greatest energy savings and keep your home more comfortable, especially in late afternoon. Experts usually suggest that you avoid shade trees on the south side, and instead use awnings, screens, and other exterior shades to block the sun coming from the south. This will give you a better wintertime heating benefit. You should also consider planting trees around light-colored gravel and asphalt surfaces which would otherwise reflect glare and radiate heat into your home. Small trees and shrubs planted close to the air-conditioner compressor will improve its performance during the cooling season, as long as the air flow isn't blocked.

Another technique is to use foundation planting to shade the lower wall areas of the home. This helps keep the ground next to the house cool and blocks reradiation from adjacent hot surfaces. You should contact your county or state agriculture or extension service to find out what plants, shrubs, and trees are recommended for your location, and get information on planting and maintaining them. Also find out what trees produce fruits for birds, squirrels, and other wildlife. They can add an extra special touch to your landscaping.

Here's a tip from landscaping experts. When planting trees to shade the walls of your home, plant them anywhere from seven to twenty feet from the house (the larger the tree, the further from the house). Research shows that the shade from trees planted ten feet from a house will last four times as long as shade from a tree twenty feet away, since the shadow from the closer tree moves across the house much more slowly.

One final thought concerns grass and ground cover. Tremendous amounts of energy and water are used each year to maintain lawns. Some estimates say that as much as 95 percent of the energy used in maintaining lawns goes for fertilizers, pesticides, and water—only 5 percent is for gasoline for lawn mowers. Choosing the right type of grass or ground cover can minimize energy expenses while keeping your lawn attractive.

Passive heating

The basic principle behind passive solar heating involves large areas of glass on the south side of a building to allow sunlight to enter. The solar heat is absorbed and stored in thick masonry walls and floors or in special liquid-filled containers. The natural movement of heat—through conduction, convection, and radiation—carries this heat throughout the living space. (See chapter 11 for detailed explanations of these heat movement principles.)

The concept of using the sun for warmth is not a new one. The writings of Socrates more than 2,000 years ago show mention of passive solar heating principles. Open porches

A passive solar home in Massachusetts (*Acorn Structures, Inc.*).

for shading and heat collection appear on Greek ruins that date back thousands of years. In modern times, the passively heated home shown at the Chicago Century of Progress Exposition in 1933 and 1934 stimulated considerable interest in this building strategy. The successful demonstration by George and Fred Keck led many people to build passively heated houses.

Since then, builders throughout the northern United States have sought to use building materials, insulation, windows, and other elements of a home to capture and use the sun's warmth. While passive cooling strategies call for plans to block the sun, heating strategies seek ways to allow it into the house, trap and store it for later use, and direct it throughout the home.

Five distinct elements are involved in passive heating systems to maximize their effectiveness:

- Collector: The area through which sunlight enters the building. It is a large glass or plastic area facing south, and is unshaded by other objects from 9 A.M. to 3 P.M. during the heating season.

- Absorber: Some type of hard darkened surface is located in the direct path of the sunlight. This is usually a masonry wall or floor, but it can also be a storage drum located especially for the purpose of absorbing the heat.
- Storage unit: The term "thermal mass" is often used to refer to thick masonry such as concrete blocks or bricks that retain the heat absorbed from the sunlight. The storage elements are usually located behind or below the absorber materials.
- Distribution system: Passive designs usually use the natural heat transfer modes (conduction, convection, and radiation) to circulate heat from the collector and storage areas to the rest of the house. Where this system is not adequate for the job, fans, ducts, and blowers can be used to help with distribution.
- Control device: Some type of movable insulation is used to prevent nighttime heat loss from the collector area. Thermostats, operable vents, roof overhangs, and awnings are among the other control devices often used in passive heating systems. The performance of the entire system depends on this control device.

How well can these principles work? Consider the case of two engineers at the Solar Energy Research Institute, who built a home in Boulder, Colorado, in 1985 that incorporated several passive solar principles, including a high level of insulation, a large expanse of south-facing glass, and energy-efficient windows, as well as an active solar water heating system.

They have closely monitored the performance of the 1,700-square-foot home since it was finished in 1986, and report that their monthly energy use is about 10 percent of that of a typical new house (their average monthly bill: $15), while their winter energy bill in Colorado's cold climate is only *2 percent* that of a typical house: $3 per month for heating in winter! Thanks to the passive home design, they're enjoying their comfortable home, no matter how cold it gets outside, with almost no energy costs.

Direct Gain

The simplest passive heating strategy is direct gain, which involves leaving the south side of a home unshaded in winter to allow sunlight into the house, where the solar heat can be absorbed and stored by the walls and floors. The home is built with large areas of windows facing south, and overhangs to block the higher summer sun but still allow the lower winter sun to reach the interior of the home.

Thick masonry walls and floors absorb the solar heat, where it can be held for several hours, depending on the climate. Window coverings are used to prevent overheating and to keep heat from escaping at night.

One other idea is to have a brick chimney located inside the home where the sun can reach it. The chimney itself will then serve as a form of thermal mass to hold the heat inside the house.

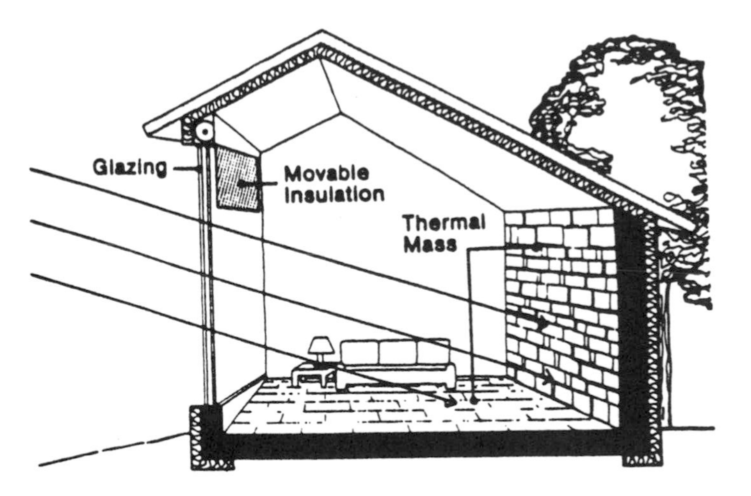

Direct gain through a south-facing window (*U.S. Department of Energy*).

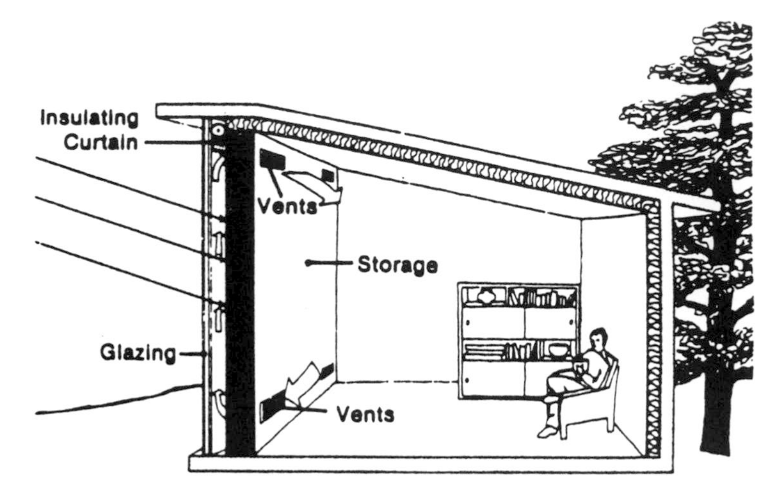

A masonry (Trombe) wall (*U.S. Department of Energy*).

Masonry Walls

A heavy masonry wall (often called a Trombe wall), made of concrete, brick, block, or other heavy material, can be built in the home to absorb sunlight. These walls are either inside or outside of rooms with south-facing windows, with a narrow air space separating them from the glass.

An alternative is to use containers such as drums or barrels filled with water to act as a thermal storage wall. One home built in the 1970s in New Mexico, in fact, used 2,000 wine bottles filled with water which were laid on racks between two glass walls.

In general, masonry is preferred because it stores heat much longer than water, even though the water will collect more heat than the same amount of masonry. The choice will depend on the climate and the heating needs. In general, heat can be stored in walls for as long as twelve hours in mild climates, or half that in colder climates.

The heat stored by the wall is radiated into the living areas of the home, and warm air in the heated space rises over the top of the wall into the rest of the house. Cooler air near the floor is drawn into the air space and heated.

Insulation

Depending on where you live in the United States, local builders insulate homes to certain R-levels. The term R-value is used to measure the insulation's ability to resist heat flow (out of your house in the winter, into it in summer). The higher the R-value, the greater the resistance of the material. These can range anywhere from R-30 in ceilings in the central part of the country to R-60 or R-70 or higher in the northern regions. Recommended wall R-levels also vary considerably depending on climate.

The type, thickness, and density of the insulation material determines its R-value. And it is important to note that R-values are additive—more insulation increases the R-value to a higher level.

One of the leading causes of wasted energy in houses is inadequate insulation. Yet it is relatively easy to add insulation to almost any house. If your home has structural framing exposed, such as in an unfinished attic or an unheated area, you can probably install insulation yourself. In other conditions, a qualified contractor can install the best type of insulation for your needs. Depending on the climate and other general variables, insulating your home to adequate levels should pay for itself in fuel savings.

A properly insulated home uses much less energy in the heating season than does a home with minimal or no insulation. Insulation also helps keep the house comfortable and reduces noise from the outside.

Because heat flows from a warmer area to a cooler area, heated air from inside your home will be lost to the outdoors in winter if insulation is not used to keep it inside.

There are several different kinds of insulation widely used in homes, including blankets (rolls of insulation), batts (blankets precut into standard sizes), blown-in (usually rock wool, fiberglass, or cellulose fiber particles sprayed into place), and foam (either liquid or rigid boards). The kind you use will depend on how much insulation is needed, where it is to be applied, availability of different types, and the cost.

If you decide to add insulation by yourself, check with a local hardware store for safety instructions. You need to be especially careful when working with insulating materials because they can be irritating to the skin and lungs. In addition, it can be difficult working in a confined attic space, so think carefully about the job before attempting to do it yourself.

Before hiring an insulation contractor, get cost estimates from at least three contractors. Find out in advance what the R-values will be, and what types of insulation will be used. It is fairly easy to measure your attic and walls to see how much insulation you now have. It doesn't pay to overinsulate, so the best job will add to your present insulation to bring you to levels recommended for your climate.

Window Glass

Not too long ago, the typical home window was a single pane of clear, untreated glass with an R-value of less than 1. Yet, just recently, a quadruple-paned, gas-filled window with low-emissivity coatings and an R-8 rating was introduced to the market. Researchers are working on even more improvements which are expected to bring window ratings to R-10 or even higher. Maybe no other area of home design has changed so much in recent years as have the types of windows available today.

It has been estimated that as much as 5 percent of the energy used in homes is lost through windows. In fact, one researcher says that the heat leaking through windows equals the annual output of the Alaska oil pipeline!

Passive solar designs, though, are based on the principle of large windows on the south-facing side of the house. And people like windows—they want to take advantage of the view from their home, they want to let in sunlight to brighten the inside, and they like the look of windows in homes. So here is an area where technology can play an important role in reducing lost energy.

In addition to double- and triple-pane windows, many

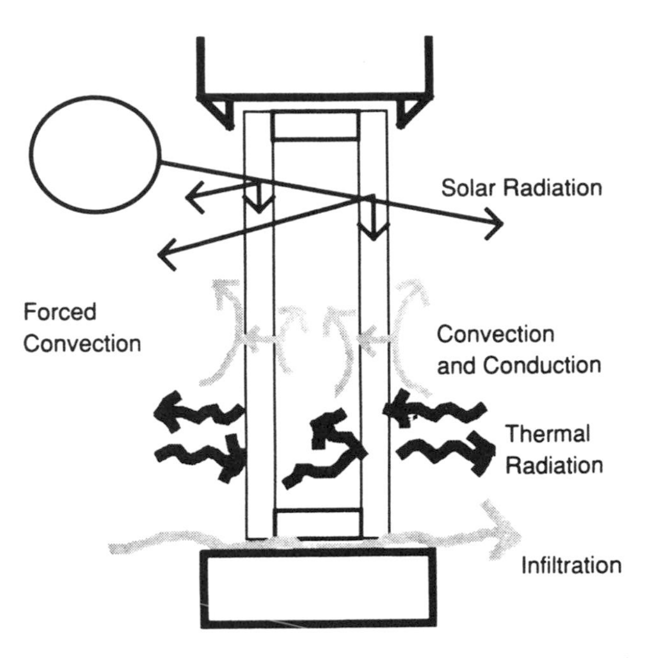

Modes of heat transfer through a window (*New Mexico Energy, Minerals and Natural Resources Dept.*).

manufacturers now offer windows that incorporate relatively new technologies aimed at increasing R-values. These technologies include low-emissivity or "low-e" coatings, and gas-filled spaces between panes. It's been less than a decade since manufacturers introduced the first low-e window, and it is estimated that half of the homes built today already are using this special glass.

Low-e coatings are microscopic thin-film metal or metal-oxide layers which are barely visible—the coating is only a few hundred atoms thick, so it looks invisible to the eye. The low-e coating may be on one or more of the glass pane surfaces or may be a thin plastic film between the window panes. The coating acts to suppress the flow of radiant heat between the glass by reflecting it back into the home during cold weather and back outdoors during hot weather. This significantly increases the R-value of the glass.

Table 8-1. Comparison of window R-values (*New Mexico Energy, Minerals and Natural Resources Dept.*).

	R-Value (hr–ft²–F/Btu)			
	GLASS ONLY	INCLUDING FRAME		
Glass Type	**Center**	**Edge**	**Aluminum w/o Thermal Break**	**Wood**
Single glass	0.90	(n/a)	0.76	1.11
Double glass, 1/2 in. air space	2.04	1.69	1.15	2.04
Double glass, E* = 0.40, 1/2 in. air space	2.44	1.85	1.23	2.33
Double glass, E = 0.15, 1/2 in. air space	2.94	2.00	1.32	2.56
Double glass, E = 0.15, 1/2 in. argon space	3.57	2.13	1.37	2.78
Triple glass, 1/4 in. air spaces	2.63	1.92	1.27	2.38
Triple glass, E = 0.15 one pane, 1/4 in. air spaces	3.03	2.04	1.32	2.56
Triple glass, E = 0.15 two panes, 1/4 in. air spaces	3.57	2.13	1.37	2.78
Triple glass, E = 0.15 two panes, 1/2 in. argon spaces	6.67	2.33	1.54	3.45

Source: WINDOW 3.1, a computer program for calculating the thermal and optical properties of windows. Based on ASHRAE standard winter conditions: 70°F indoor air temperature, 0°F outdoor air temperature, 15 mph outdoor air velocity, and no incident solar radiation.
*"E" is the emissivity of the low-E coated surface.

Many hardware stores now sell low-e stick-on window films that are very cost-effective.

Other types of windows are being tested and introduced to the market. Spaces between double- and triple-pane glass can be filled with gases which insulate better than air. Argon sulfur hexafluoride and carbon dioxide are the main gases used commercially today. Gas-filling adds barely any cost to the glass and provides added R-values when used with low-e coatings. If all the single-pane windows are replaced, double-pane, gas-filled windows with low-e glass can save from $35 to $65 per month, depending on the local energy costs and R-value of the windows.

Researchers are also looking at the use of aerogel, an unusual solid material that is both porous and transparent and contains insulating air cavities. One-half inch of aerogel in a double-glazed window might increase the window's R-value by 500 percent! Also under study is glass that insulates against heat loss but has the capability of changing its own properties to respond to the home's need for light, shade, or warmth. These "electrochromic" glazings basically allow windows to be turned on or off to control the transmission of light or heat. This system has already been demonstrated on a Japanese automobile with an electrochromic sunroof that can be adjusted to allow from 5 percent to 30 percent of the sunlight striking it to enter the car.

Sunspaces

One of the most popular types of passive heating elements used in the northern U.S. today is the attached sunspace. This type of room, which is used as a home heating system, year-round garden, or an extra living space, combines the best features of both the direct gain and the thermal storage wall building techniques. Research has shown that sunspaces can provide as much as 60 percent of a home's heating requirements in winter depending on the climate, the size of the sunspace, and the heating use.

These rooms, also called greenhouses or sunrooms, are built onto the south wall of the home. Heat comes through the south-facing glass and is stored in a thermal wall, though thick floors, benches, and tanks of water can also be used for storage. At night, when the temperature in the sunspace drops, the stored heat is slowly released by the storage mass. Windows and vents in the storage wall allow this heat to be channeled into the home's indoor living space. In the summer, the sunspace can help with cooling by allowing heat to escape through its vents, and the storage wall absorbs heat before it can get into the house.

Homeowners have some choices when deciding how the sunspace will be joined to the rest of the house. One method is to have an uninsulated brick or concrete wall separate it from the home. The wall will absorb and store heat from the sun. Another idea is to have large windows and sliding glass doors where the sunspace is joined to the house. In this case,

An attached sunspace (*U.S. Department of Energy*).

A sunspace creates an enjoyable extension of the home's living area (*Acorn Structures, Inc.*).

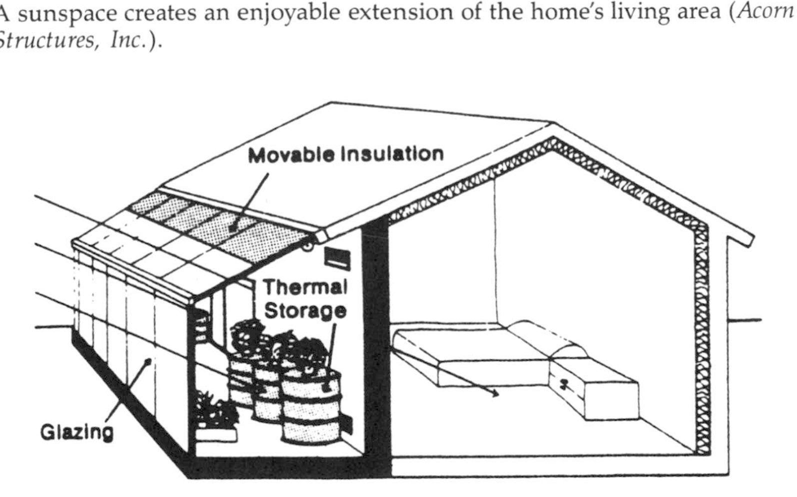

Attached sunspaces can keep a home much warmer in winter. A detailed drawing is shown on page 134 (*U.S. Department of Energy*).

a masonry floor would act as the main heat storage feature. This floor needs to be left uncarpeted since coverings such as carpet prevent the heat from being absorbed.

Sunspaces also need some kind of material to cut down on nighttime heat loss through the glass. Movable insulation that the occupants can control is one option. Another choice is special glazing, mentioned earlier, that will help reduce heat losses.

Finally, you should note that growing plants in a sunspace can cut down on the heat output. Depending on the

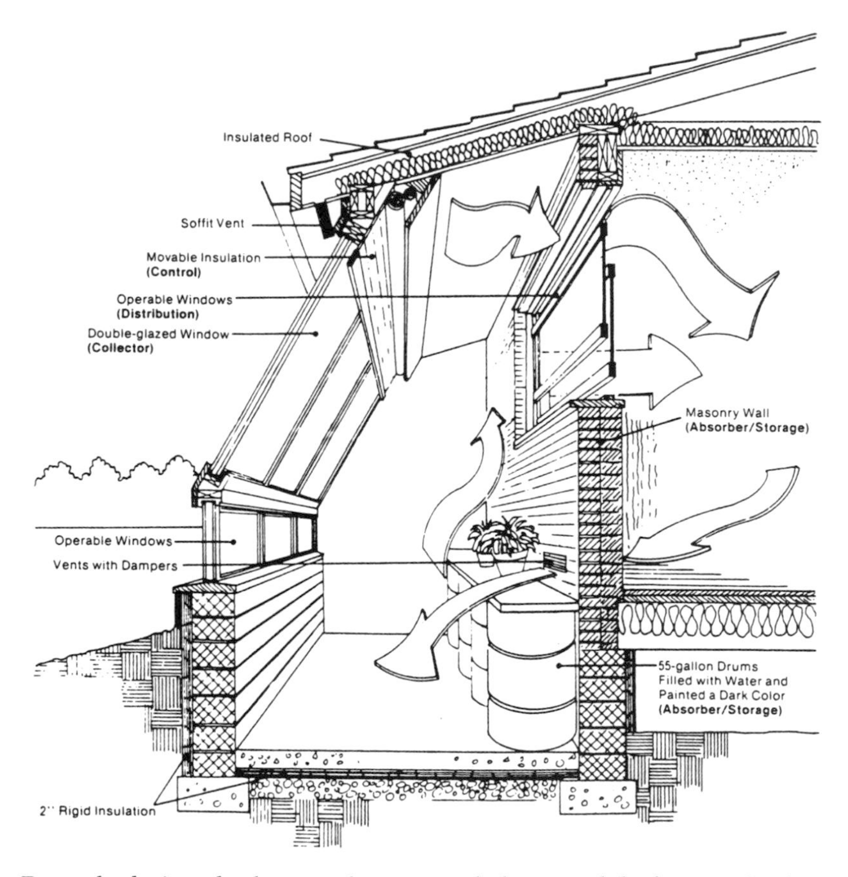

Properly designed solar greenhouses can help control the heat getting into a house (*U.S. Department of Energy*).

type of plants you grow, evaporation from the leaves can reduce the sunspace's heat gain by as much as 50 percent. On the other hand, plants do act as natural humidifiers and can add to the home's comfort in dry regions of the country, but use of common plant pesticides must be reduced since they will be vented into the house.

Landscaping

You can use landscaping to help insulate the home, block cold northern wind, and collect solar radiation. In cold climates, a row of trees wider than it is tall, planted perpendicular to the wind, can act as a windbreak and help keep the house more comfortable. Ideally, such a barrier should be planted along the northwestern side of the house to block the winter wind.

To allow maximum winter sun and some summer shade, a good planting strategy for cold climates is to locate deciduous trees at a distance three times their expected height from the building. This means that a tree expected to grow twenty-five feet tall should be planted seventy-five feet from the house.

Other planting strategies include screening ponds and other small bodies of water with a row of shrubs or trees to deflect the winter wind, and using earthberms to store and radiate heat into the house. These earthberms help deflect wind over the roof. Carefully placed hedges can block winds from reaching doors and garages, and can even blow snow away from these areas.

It's usually best to use indigenous plants whenever possible, so contact your local garden club or nursery for information on trees and plants best suited for your climate. Among the deciduous trees best suited for cool climates are silver maple, horse chestnut, birch, oak, and crab apple, while coniferous trees include balsam fir, spruce, hemlock, Scotch pine, and Douglas fir.

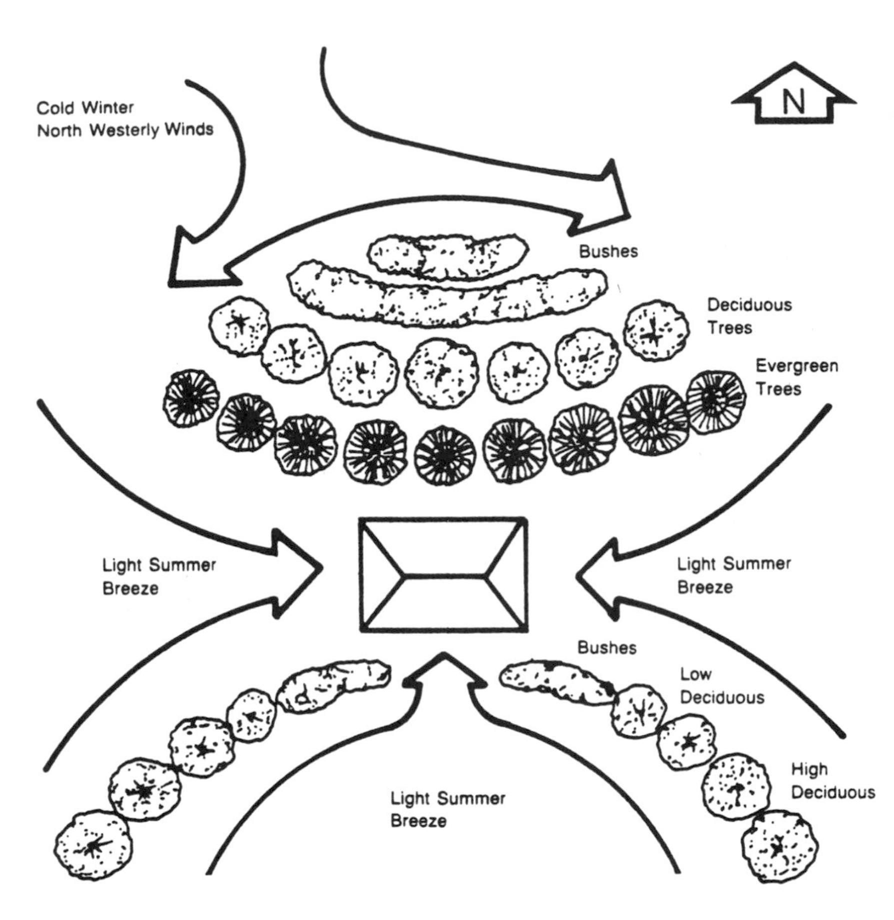

The best windbreaks will block winter winds and funnel summer breezes (*U.S. Department of Energy*).

Research shows that a carefully placed row of evergreens can lower your heating bill by as much as 20 percent. Don't neglect landscaping as a tool to use in lowering energy use while improving indoor home comfort.

CHAPTER NINE

Financing and Economics

How to buy the right system to fit your pocketbook

There are basically two ways to buy solar water heating systems today—either as part of a new home with the cost of the system included in the mortgage payments, or as an add-on to an existing home in which you pay the cost outright to the contractor, usually through some type of financing plan set up with a local bank.

In either case, the monthly payments are almost always less than the energy savings, paying back your investment from the very start.

Consumers interested in buying photovoltaic (PV) products today, however, usually just pay the money directly out of pocket. Most of the solar electric items available at this time are still in the small product and novelty stage, and are bought off-the-shelf in hardware stores, specialty shops, and catalog outlets. They're relatively low-priced, easy to install, and usually used for small needs around the home—auto battery chargers, lighted mailbox numbers, music boxes, moving sculptures, and key chains.

Larger photovoltaic products gaining popularity in the marketplace include patio and walkway lights, mole and rodent evictors, and attic vent fans. Even these bigger products are easily affordable by most homeowners, and they provide either an energy-saving or labor-saving service, or both.

While consumers in most cities can check the yellow pages of their phone book for solar water heating contractors and companies, and photovoltaic dealers and distributors, it may still be difficult to find a wide selection of products to choose from in some parts of the country. However, there are a number of national mail order electronics, gardening, and lifestyle catalog outlets that feature solar products. As these products gain in popularity, more items will be available from the catalogs.

The following mail order firms or organizations currently include popular solar products and larger solar systems in their catalogs, and can be contacted for more information:

- Edmund Scientific (for solar cells and small solar motors), 101 East Gloucester Pike, Barrington, NJ 08007-1380.
- Home Power, P.O. Box 130, Hornbrook, CA 96044.
- Jade Mountain, P.O. Box 4616, Boulder, CO 80306.
- Real Goods News, 966 Mazzoni Street, Ukiah, CA 95482.
- Seventh Generation, Colchester, VT 05446-1672.
- The Solar Electric Catalogue, 175 Cascade Court, Rohnert Park, CA 94928.

These companies, along with other variety stores in your town, offer the popular battery chargers, flashlights, and other devices that are widely used by consumers today. But beyond these simple gadgets, there are many larger items that can make dramatic improvements in your lifestyle, significantly lower your energy usage, and otherwise allow you to take advantage of solar energy in many facets of your lifestyle today.

To decide if a solar system will help you today, you need to ask three basic questions:

1. Is reliability an important requirement, and, if so, what times or circumstances?
2. What are the installation, fuel or storage, and maintenance costs of the alternatives?
3. When will the initial solar investment be paid back, and have I taken advantage of any special local, state, or federal solar incentives that may apply?

The following analysis is divided between solar electric applications and solar water heating options. Note that in each case, the same basic issues apply, but each has different considerations that deserve unique attention.

Solar electric applications

Two basic types of available solar applications are considered for your home in this section—yard uses (solar patio or walk and driveway lighting), and home uses (solar security lights or attic fans).

In any solar application, the key ingredient is to consider when the lighting or ventilation is needed and if there is solar access to provide it. For example, if you will only need the lighting or ventilation for a few hours in the evening rather than all night, solar electric power can be an ideal option. To be most effective, though, the solar panel must be totally unobstructed so the maximum amount of sunlight will strike it. In most cases, PV panels are mounted atop a light pole or as part of the walk/driveway light. Take a look at the site where the panels will be and see if any trees, buildings, or other objects will block out sunlight at any time during the day (typically from 9 A.M. to 3 P.M.). If there will be some obstruction, you may wish to consider relocating the light or appliance with the solar panels to another area where it will receive the greatest amount of sun, or arrange to have the panels located in this open area and wired to the appliance. Before you make the final decision on locating outdoor solar lighting, it is essential that you spot the sun during one full day to assure direct solar access during the peak solar times. Shadows can drastically cut down on the performance of the cells.

Reliability of the appliance is affected by the performance of the solar cells, so most solar contractors use special tools to measure the available sunlight and locate potential obstructions before the product is installed. Walk/driveway lights, for example, need a strong charge during the day to perform their job at night. Cutting down on that sunlight will diminish the performance of these products.

The second issue involves economics relating to the alternatives you can choose from to do the required job. In most cases, the alternative to using photovoltaics is to run electric wires to the light. Close to the home, there may be electric outlets that will allow the conventional lighting product to be simply plugged in and used. While this will obviously increase your energy use, it can be very convenient and easy to do.

However, there are important issues regarding safety and variables of the location which can often make the use of electricity impossible or even dangerous. In terms of sheer economics, many costs can make outside uses of electricity prohibitive. Consider a situation where you want to put lights along a walkway in your backyard. Until the advent of solar photovoltaics, the option for the average homeowner was to hire an electrician to run electric wires to the site. As an estimate, let's say that the minimum costs of the electrician will be $50 for the first hour of labor and $10 a foot for running the wire. So if you're running an underground electric line for thirty feet, for example, to provide power for the walklights, the cost would be $350 for the electrical work, in addition to the cost of the light itself. In most cases, a solar patio light or walk/driveway light would cost much less than the installation and purchase price of a traditional electric lighting system. All you have to do to install the solar product is stick it into the ground or hang it where you want it.

But that's obviously not the entire economics issue. You've got to consider the expenses of electricity to run the conventional electric light. While an average light might cost only around $1 or $2 per month for electricity during the year—not a lot of money when you look at it that way—it does add up to another $60 to $120 in costs in just five years, assuming there is no big increase in electric rates which would make the cost even higher. Then add to this the fact that conventional light bulbs burn out much faster than do lower-energy photovoltaic bulbs—an issue to consider for patio lights which use standard bulbs (but not necessarily a problem for electric walkway lights which often use special low-voltage bulbs).

As a general rule-of-thumb, you can figure that for outside uses like walk/driveway lights, solar electric lighting products are cost-effective when compared with electric alternatives if an electrician must be used for the installation and wire needs to be run more than fifteen feet from the electric source. If you're a do-it-yourselfer and don't need to hire someone to do the work, you ought to figure in the value of your time and effort to get a better idea of what the true cost of installation is for the product. Economists use estimates like these to do life-cycle calculations, in which the total cost of a product, including purchase price, installation costs, and energy use, are figured to make accurate comparisons of energy alternatives.

When looking at alternatives to electrical products for indoor uses, these same considerations apply. In addition, though, one other factor needs to be considered. Many states and communities have strict building codes that regulate the safety of electrical products used in the home. Any time electrical work is done in the house, you need to be sure that you are meeting required codes and regulations.

For example, a very popular use of solar energy today is

to power attic vent fans to keep the home more comfortable, especially during the hot summer months. However, an electrically powered vent fan can use one dollar's worth of electricity every day in areas where time-of-day metering is in effect (a trend in many utility companies' rates these days). But a solar-powered vent fan will not only lower cooling bills by 5 percent, it won't need any electrical energy for power, either. So the solar option makes the economics very good for uses like this. Note that cost considerations regarding the economics of photovoltaics are discussed in chapter 5.

This is where a close look at local ordinances is essential. For example, it is imperative that wall penetrations are made in accordance with local building codes to ensure against leaks. Running wires from the outside solar panels to the inside fan makes an opening in the shell of the house, which must be properly sealed. Other codes and regulations affect electrical products as well. Because of this, a number of states have adopted building codes regulating the equipment, installation, and warranties of solar products. See the state-by-state listing later in this chapter that outlines current regulations and solar building codes that must be adhered to by the homeowner or installer.

For payback estimates for solar products, see the chapters on solar water heating, pool heating, and photovoltaics. They offer guidelines to use in determining if the system will be cost-effective for your needs.

Solar heating applications

At the present time in most parts of the country, solar space heating is more expensive than heating your home with natural gas. The use of solar energy for home heating may be economical, though, in areas of the country where electricity is used for heating, and where utility costs exceed fifteen cents per kilowatt hour.

There are certain specific applications where the solar option may be the best economic alternative. For example, assume that you are interested in heating a room such as a recently enclosed porch which has no heating system in it at the present time and is used during the daytime. An economic option for this case would be a solar air-heating system mounted either on the roof of the porch or attached to a southern wall of the home. The energy need is during the day (when the sun is shining), and the expense of installing an electric baseboard heating system or extending the home's heating system to the porch will clearly cost more than will the solar space heater. Because the porch is used during the daytime and heat storage is not an issue, all you will need is the solar collector and a vent fan to direct the hot air. Solar space heating might also be economic in homes that are not attached to the utility grid. These remote homes, weekend retreats, and vacation cabins are often in areas where extension of the electric lines or the use of propane heating systems are not practical or affordable.

There are several basic economic issues that will determine whether or not a solar system will be cost-effective for heating either water or air in your home:

- Is conventional energy readily available? If it is not easily obtainable (you may be located too far from the power lines, for example), and if there is adequate solar access (shade-free space to locate the collectors), then a solar space heating system can be considered.
- If conventional energy is available to the site, but you are using it mostly during the midday, or if electric rates are higher during the midday in your area, solar heating can be a good alternative to conventional heating systems.
- If regular supplies of fuel oil or propane are not reliable or if you plan to live in the house for many years and want to assure your family the availability of free energy in the future, you can consider solar heating systems.

For specific economics of systems, consult chapter 10.

Financing all kinds of solar systems

Both solar energy heating systems and photovoltaic systems can be financed through FHA/VA government mortgage programs as well as private financing options such as home improvement and retrofit loans. In order to receive loan approval, you will need to answer the questions given earlier in this chapter regarding life-cycle costs and alternatives to paint a lifestyle and economic picture as to why you are making this solar purchase.

You also need to run the basic economic calculations showing the costs and system paybacks, which are explained fully in the appropriate chapters of this book.

Your proposed installation contractor will have to show that he is properly certified to install solar systems in the county or other locality and that the system will meet the appropriate local building codes. Prior experience in installing solar systems by your contractor is usually essential for approval by the financing entities.

The SEIA has conducted a survey on the extent of state involvement in regulating solar companies relative to certifying and licensing installers, equipment, warranty programs, and building code enforcement programs (table 9-1).

In regards to solar installer certification programs,

Table 9-1. Current state solar regulations (*Solar Energy Industries Association*).

	IN	EQ	WA	CE		IN	EQ	WA	CE
Alabama	□	○	○	□	Georgia	□	□	○	□
Alaska	○	○	○	○	Hawaii	□	○	○	□
Arkansas	●	●	○	●	Idaho	○	■	■	□
Arizona	□	□	□	□	Illinois	○	○	○	○
California	●	○	○	○	Indiana	○	□	○	○
Colorado	○	○	○	○	Iowa	○	○	○	□
Connecticut	●	○	○	□	Kansas	○	○	○	○
Delaware	□	○	○	□	Kentucky	○	○	○	○
District of Columbia	□	■	○	□	Louisiana	○	○	○	○
Florida	●	●	○	○	Maine	●	○	○	○

	I N	E Q	W A	C E
Maryland	□	○	○	□
Massachusetts	○	○	○	○
Michigan	□	□	■	●
Minnesota	○	○	○	□
Mississippi	○	○	○	○
Missouri	○	○	○	○
Montana	○	○	○	○
Nebraska	○	○	○	○
Nevada	●	○	○	○
New Hampshire	○	○	□	○
New Jersey	○	□	○	○
New Mexico	□	□	○	○
New York	○	○	○	○
North Carolina	□	○	○	○
North Dakota	○	○	○	○
Ohio	○	○	○	○
Oklahoma	○	○	○	○
Oregon	□	□	○	○
Pennsylvania	○	○	○	○
Rhode Island	●	○	○	○
South Carolina	□	□	○	○
South Dakota	○	○	○	○
Tennessee	○	○	○	□
Texas	○	□	○	○
Utah	□	□	□	○
Vermont	○	○	○	○
Virginia	□	○	○	□
Washington	□	○	○	○
West Virginia	○	○	○	○
Wisconsin	○	○	○	○
Wyoming	□	○	○	○

KEY

INSTALLERS

○ No contractors license required.
□ Installer must be licensed contractor, but no solar specific license is required.
● Installer must be certified/licensed in solar installations.

EQUIPMENT

○ No certification required.
■ State certification in place on a voluntary basis.
□ Certification required to qualify for any state tax incentives, loans, etc.
● Certification required to market systems.

WARRANTY

○ No programs.
■ Voluntary programs.
□ Minimum requirements to qualify for tax incentives, loans, etc.

CODE ENFORCEMENT

○ None on the state level.
■ Model code on the state level is adoptable on the local level.
□ State requires permit approval and post construction inspection as for all mechanical/electrical installations.
● Specific solar codes in place for installation inspections.

twenty-three states currently require some type of contractor license before approval for installation will be given. Seven of these states—Arkansas, California, Connecticut, Florida, Maine, Nevada, and Rhode Island—have legislation that specifically calls for a solar license category or solar installer certification. The other states do not require a license or certification of any kind, though many local governments do have their own regulations that cover solar licensing requirements.

Some type of equipment certification program is in effect in fourteen states, with ten of them requiring certification to qualify for tax incentives, loan programs, or other special financing opportunities.

Not many states have solar installation warranty programs. Only Arizona, New Hampshire, and Utah have minimum warranty requirements, and this is necessary to qualify for tax incentives and loan programs. Both Idaho and Michigan have voluntary programs that recommend minimum warranty standards, but these cannot be enforced by law.

Finally, fifteen states report that they have a building code enforcement policy. Specific building codes are in place for installation inspections in Arkansas and Michigan. The other thirteen states use the standard electrical, mechanical, and plumbing codes that can be applied to solar installations. The basic enforcement in these fifteen states involves construction permit approval and postconstruction inspections.

To determine if there are other regulations that affect solar installation in your state, or if there have been updates, get in touch with your state solar contact listed in chapter 10.

Solar water heaters must also have the approval of the Solar Rating & Certification Corporation (SRCC). For pool and space heating systems, the solar collectors should be SRCC-certified. Most photovoltaic panels have the UL label, as do the other electrical components in the system.

As mentioned earlier, the FHA and VA loan programs have incentives for energy-efficient expenditures for new homes. When buying a new house, these can be incorporated into the loan, and FHA/VA may increase your debt-to-income ratio, allowing you to afford a higher-cost home because the building maintenance expenses will be lower. Aside from SRCC certification on solar water heaters which can give automatic approval to incorporating the costs in the loan, other systems will have to be reviewed by local FHA offices to ensure that they comply with the HUD Minimum Property Standards. All conventional solar systems installed by an experienced contractor should qualify for these programs.

Because solar energy is relatively new, the FHA or a private bank will want to be sure that you are paying a fair price for the solar system. The easiest way to prove this is an installer's picture book with prices of his most recently installed systems, or recent promotional brochures or fact sheets on other comparable solar systems. Most installers use these materials as a regular part of their business. Letters from individuals with a comparable solar system in your area can also be sufficient to meet this requirement. The idea is to provide the bank or lender with a good idea of the worth of the system. Your installer can provide the paperwork needed for larger photovoltaic and solar space heating applications.

In retrofit loans from private banks, known as "home improvement loans," you usually will not need to provide collateral for a system costing less than $2,500. Solar systems above this price may require collateral or a down payment of 20 percent of the system cost.

When looking at the economics and financing aspects of a solar system, you can readily see what makes this type of purchase unique. A solar system is one of the few—and maybe the only—items you will ever buy that will actually pay for itself during its use. Look at this example to see how it applies to a retrofit home improvement loan.

Consider a solar water heating system that costs $2,200 installed, and you pay for it with a five-year loan financed at 12 percent. The cost of this loan would be $48.94 per month for sixty months. But to calculate what it really costs in terms of energy use, you need to subtract the monthly loan payments from the monthly utility savings.

Let's assume that this system saved you $45 per month in energy costs during the first year. Thus, you would be paying only $3.94 per month out of pocket during the first year for your solar water heater. Assuming just a 3 percent annual increase in energy costs, you would be saving $49.17 per month by the fourth year—more than enough to pay your full monthly payment. At the end of the fifth year, the system would be paid for, and the monthly savings would be cash in your pocket—thousands of dollars in savings over the twenty-year lifetime of the system.

Since modern solar systems have only been widely available in mass scale during the past decade, most federal, state, and private financing institutions are only now beginning to become familiar with solar energy and its benefits. In states where solar has been in widespread use for many years, such as California, Arizona, New Mexico, Florida, New Jersey, Colorado, and some others, there should be little trouble getting the right kind of financing for your systems. In areas where solar is less known, more substantiation of savings and system reliability along with professional installation will make banks more assured of this type of investment.

Assuming your credit rating is acceptable to the bank, the solar energy system should readily qualify for financing. It's an easy and economical *investment*—and that's the key word to use when discussing the economics of these systems. The money spent on a solar system will return a strong dividend in terms of economic payback in a relatively short time period, making this return in most cases comparable or even greater than the return your money would yield in most types of current investments.

Whether you're planning to install a solar system in your new home or to add one to your present home, the economics make it a wise expenditure that will provide many years of more comfortable living without actually costing you any money at all.

Selecting contractors

Most of the guidelines so far in this book have covered working with solar contractors. But to incorporate these ideas in new home construction, you will need to work with architects and builders who understand solar principles and know how to fit solar systems into their plans and designs.

When interviewing architects, ask about their experience with solar systems. Find out if the architects have built homes with active or passive solar features before, whether they have attended courses or training programs on solar systems, and, generally, what their attitudes are toward using solar energy to provide heat or electricity.

Call the local architecture society or association and ask for referrals to architects who have been identified as using solar features in their homes. Passive solar designs fit especially well with many home plans and designs, and architects often find it easy to incorporate these features without sacrificing their original design or increasing your building costs. Contact the architects you learn of and ask for a list of the residential projects they have worked on which would be similar to your needs. Even if the architect does not have experience with similar homes, he or she may share your interest in using active and passive systems, so you need to discuss their philosophy on solar design.

But the best way to make the final decision is to visit some of the homes they have designed and talk with the people living there. Are the homes comfortable? Are they energy-efficient? Are they happy living in the home? Find out how they feel about this design, and what their experiences were working with the architect.

To choose a builder, you should go through similar steps to those just listed. Begin by contacting the local home builders association for a listing of builders who have experience with solar homes. You can also get referrals from real estate agents, home appraisers, banks, and others involved in the construction process.

Talk with several of the builders and get the names of some people living in homes they have built. Contact these people and see how satisfied they are with the house.

Contact your local Better Business Bureau or the state agency that regulates contractors to see if there are any complaints on the builders in their files. Local builders groups also often maintain such information on their members.

The biggest key to choosing architects, builders, electricians, landscape architects, and others involved in your project will be the experiences of other people who have worked with them. If they are satisfied with their home and feel that it meets their expectations of energy efficiency, comfort and quality, then you have an idea of what their work is like and what kind of job they may do for you.

It is usually best to choose a professional who has a business based on installing solar systems, or who has experience designing or building homes with passive features or active systems. However, you may find someone who has a strong interest in this area, has attended educational programs on solar technologies, and wants to start building this type of house. Talking philosophy and objectives can help clarify your needs and expectations, and assure that you choose the right professionals to work with on this project. The best approach is to personally meet the people referred by the solar contractor, architect, or builder to inspect the work and verify the purchaser's satisfaction.

CHAPTER TEN

Where to Go from Here

A guide to the people, organizations, and materials that can give you more information

Publications

Bibliography: Reference and Media Center. 1990. Compiled by Bill Brooks. North Carolina Solar Center (Box 7401, North Carolina State University, Raleigh, NC 27695). This is a new annotated guide to more than 700 books, periodicals, and videotapes on all aspects of solar energy. Cost: $7.

Building for the Caribbean Basin and Latin America. 1989. Compiled and Edited by Kenneth Sheinkopf. Solar Energy Industries Association (777 North Capitol Street, N.E., Suite 805, Washington, D.C. 20002). This book covers all aspects of building an energy-efficient home in hot, humid climates, including how to choose solar water heating equipment. Cost: $35 plus $3 postage and handling.

Catalog of Renewable Energy Publications. 1991. Solar Energy Industries Association (777 North Capitol Street, N.E., Suite 805, Washington, D.C. 20002). This catalog describes dozens of publications available from the major national research laboratories and educational organizations. Cost: Free.

Cooling with Ventilation. 1986. Subrato Chandra, Philip Fairey, and Michael Houston. Solar Energy Research Institute (1617 Cole Blvd., Golden, CO 80401). This detailed manual covers such subjects as window shading, radiant barriers, air circulation fans, window design, and naturally ventilated home designs. For availability and cost, contact the Superintendent of Documents, U.S. Government Printing Office, Washington, D.C. 20402.

Directory of the U.S. Photovoltaic Industry. 1990. Solar Energy Industries Association (777 North Capitol Street, N.E., Suite 805, Washington, D.C. 20002). This publication includes project descriptions of telecommunications, lighting, water pumping and treatment, navigational and transportation systems. All PV manufacturers, distributors, dealers, and component manufacturers in the U.S. are listed. Cost: $15 plus $2.50 postage and handling.

A Golden Thread. 1980. Ken Butti and John Perlin. Cheshire Books (Palo Alto, CA). This is a comprehensive history on the development of solar energy. Many excellent pictures dramatically show the advances in the technology.

A Guide to the Photovoltaic Revolution. 1985. Paul D. Maycock and Edward N. Stirewalt. Rodale Press (Emmaus, PA). An excellent overview of the photovoltaic process, with many examples of applications and uses of the technology.

Solar Thermal: A Directory of the U.S. Solar Thermal Industry. 1989. Solar Energy Industries Association (777 North Capitol Street, N.E., Suite 805, Washington, D.C. 20002). This directory includes detailed product information from U.S. solar thermal manufacturers and engineering/design companies. It gives names and addresses of manufacturers, distributors, and others involved in the solar water heating, pool heating, and other thermal products industry. Cost: $10.00 plus $2.50 postage.

Solar Water Heating: A Consumer Guide. 1990. Florida Solar Energy Center (Public Information Office, 300 State Road 401, Cape Canaveral, FL 32920). This booklet offers consumers information on comparison shopping, warranties, economics, financing, and related topics. Cost: Free.

Solar Water Heating—Is It For You? Undated. Northeast Utilities (P.O. Box 270, Hartford, CT 06141-0270). This guidebook explains how solar water heating systems work, what to look for when buying a system, and understanding the economics of solar water heating. Other publications from Northeast Utilities include *Protecting Solar Access: Tips for Builders, Homeowners, Developers and Site Planners; Regulating Passive Solar Subdivision Design; Passive Solar Living,* and *The Solar Home Plan Book.* Write for cost and availability of these publications and others in their "Operation SOLAR" series.

Organizations

American Council for an Energy-Efficient Economy (1001 Connecticut Avenue, N.W., Suite 535, Washington, D.C. 20036. 202-429-8873). Provides information on energy efficiency of home appliances and other conservation products. Their annual edition of "The Most Energy-Efficient Appliances" lists the top-rated residential air conditioners, furnaces, water heaters, heat pumps, and refrigerators by brand name and model number. Copies of the current report are $2 each.

American Solar Energy Society (2400 Central Avenue, B-1, Boulder, CO 80301. 303-443-3130). ASES is a national society for professionals and others involved in solar energy. It provides a variety of forums for exchange of information; publishes information of interest and benefit to the solar community, including technical and scientific information on new solar advances, and promotes education in fields related to solar energy.

CAREIRS (toll-free telephone number is 800-523-2929; 800-233-3071 in Alaska and Hawaii) is the nationwide Conservation and Renewable Energy Inquiry and Referral Service. Call them for free fact sheets and publications on dozens of different subjects in the solar energy field.

Florida Solar Energy Center (300 State Road 401, Cape Canaveral, FL 32920. 407-783-0300). FSEC is one of the largest and most active alternative energy centers in the country. Their staff conducts research on solar buildings, solar water heating systems, photovoltaics and advanced technologies, and they offer many courses and workshops on all aspects of solar energy. They also have many free publications available to the industry and the general public.

National Appropriate Technology Assistance Service (P.O. Box 2525, Butte, MT 59702-2525. Toll-free 800-428-2525; 800-428-1718 in Montana). This is a free service funded by the U.S. Department of Energy to help consumers understand and implement energy conservation and renewable energy technologies. NATAS information specialists will answer questions, discuss individual technical issues, evaluate energy projects and products, and suggest innovative financing and marketing techniques. Most NATAS inquiries deal with solar thermal energy, energy-efficient construction, energy-efficient appliances, and energy-efficient building retrofits.

National Association of Home Builders National Research Center (400 Prince George's Blvd., Upper Marlboro, MD 20772-8731. 301-249-4000). The Center is the research arm of the National Association of Home Builders and provides information to builders on new products and building strategies. Your home builder can contact them for information on energy-efficient products.

National Center for Appropriate Technology (P.O. Box 3838, Butte, MT 59702. 406-494-4572). NCAT offers a number of publications, training programs, and technical assistance efforts on energy savings for consumers.

Passive Solar Industries Council (1090 Vermont Avenue, N.W., Suite 1200, Washington, D.C. 20005. 202-371-0357). PSIC is a national association devoted to providing practical information on passive solar design and construction to the U.S. building industry. Publications, forums, and various programs provide a national network of businesses and professionals with current information on energy-efficient, high-quality buildings.

Sandia National Laboratories (Solar Energy Department, P.O. Box 5800, Albuquerque, NM 87185. 505-844-3077). Sandia is a national laboratory conducting a wide range of research activities and educational programs and services. They manage a Solar Thermal Design Assistance Center which provides design assistance and systems development of large-scale solar thermal and photovoltaic systems.

Solar Energy Industries Association (777 North Capitol Street, N.E., Suite 805, Washington, D.C. 20002. 202-408-0660). SEIA is the national trade organization of the solar thermal, passive solar, and photovoltaic manufacturers, distributors, component suppliers, and contractors.

There are currently four active state chapters of SEIA:

- California SEIA: 889 Riverside Avenue, Suite C, Roseville, CA 95678. 916-782-4809.
- Colorado SEIA: 112 East Fourteenth Avenue, Denver, CO 80301. 303-231-7673.
- Florida SEIA: 930 N. Krome Avenue, Suite 2A, Homestead, FL 33030. 305-246-8447.

- Maryland/Virginia/DC SEIA: 2201 Wisconsin Ave., N.W., Suite C130, Washington, D.C. 20007. 202-333-2749.

Solar Energy Research Institute (1617 Cole Blvd., Golden, CO 80401. 301-231-1000). SERI is a federal laboratory that conducts solar research in such areas as thermal materials, advanced solar thermal systems in buildings, and other basic and applied research areas. They also offer many educational materials for the industry and the general public.

Solar Rating & Certification Corporation (777 North Capitol Street, N.E., Suite 805, Washington, D.C. 202-408-0665). SRCC is a nonprofit national corporation which develops certification programs and rating standards for solar heating equipment. Contact them for information on obtaining a current directory of collector and system performance ratings.

U.S. Department of Energy (1000 Independence Avenue S.W., Washington, D.C. 20585. 202-586-5000). The Department of Energy produces many publications, sponsors a variety of programs and displays, and coordinates activities designed to increase the utilization of solar and renewable energy technologies in the U.S.

Periodicals

Custom Builder: The Monthly Magazine of Quality Home Construction (Box 985, Farmingdale, NY 11737). This monthly magazine provides information on many new energy-efficient building techniques and strategies. Subscription rate: $24 per year.

Home Energy Magazine (2124 Kittredge, Suite 95, Berkeley, CA 94704). Many articles on energy-saving tips and solar applications are featured in this bimonthly publication. Subscription rate: $35 per year.

Independent Energy (107 S. Central Avenue, Milaca, MN 56353. 612-983-6892). This monthly publication specializes in discussing the implications of technology and policy in power applications, markets, government, regulatory affairs, and other aspects of alternative energy technologies. Subscription rate: $72 per year.

International Solar Energy Intelligence Report (951 Pershing Drive, Silver Spring, MD 20910. 301-587-6300). This biweekly newsletter covers many aspects of the solar business, including government, industry and policy programs. Subscription rate: $365.50 per year.

Photovoltaic Insider's Report (1011 W. Colorado Blvd., Dallas, TX 75208). A monthly newsletter on worldwide developments in photovoltaics. Subscription rate: $127 per year.

Photovoltaic News (Photovoltaic Energy Systems, Inc., P.O. Box 290, Casanova, VA 22017. 703-788-9626). Each month this newsletter gives detailed information on the latest developments in the use of solar electricity, including new products and applications. A regular feature is a detailed bibliography of related books, articles and meetings. Subscription rate: $100 per year. The company also publishes the *Worldwide PV Yellow Pages,* books and other materials on photovoltaics, and four "Instant Expert" slide presentations.

Solar Industry Journal (777 North Capitol Street, N.E., Suite 805, Washington, D.C. 20002. 202-408-0660). This is a quarterly publication of the Solar Energy Industries Association, giving information on all aspects of solar energy applications and technology. Subscription rate: $25 per year.

Solar Today (American Solar Energy Society, 2400 Central Blvd., B-1, Boulder, CO 80301. 303-443-3130). In this bimonthly publication of the American Solar Energy Society, articles cover research, industry developments, and people involved in the solar technologies. Subscription rate: $25 per year.

State Energy Offices

The following list of state energy offices gives current contact names and addresses for obtaining information on solar energy programs and activities in each state:

ALABAMA: Bernard Levine, Alabama Solar Energy Center, University of Alabama at Huntsville, Huntsville, AL 35899. 205-895-6361. In-state toll-free: 800-228-5897.

ALASKA: Stuart Brooks, Department of Community and Regional Affairs, Rural Development Division, 949 E. Thirty-Sixth Avenue, Suite 403, Anchorage, AK 99508. 907-563-1955.

ARIZONA: Ray Williamson, Arizona Energy Office, Department of Commerce, 3800 N. Central, Twelfth Floor, Phoenix, AZ 85012. 601-280-1440.

ARKANSAS: Morris Jenkins, Arkansas Energy Office, One State Capital Mall, Little Rock, AR 72201. 501-682-7377.

CALIFORNIA: Alec Jenkins, California Energy Commission, 1516 Ninth Street, Sacramento, CA 95814. 916-324-3499.

COLORADO: David Warner, Office of Energy Conservation, 112 East Fourteenth Avenue, Denver, CO 80203. 303-894-2144. In-state toll-free: 800-632-6662.

CONNECTICUT: Virginia Judson, Office of Policy and Management, Energy Division, 80 Washington Street, Hartford, CT 06106. 203-566-2800.

DELAWARE: Robert Bartley, Division of Facilities Management/Energy, P.O. Box 1401, Dover, DE 19903. 302-736-5644.

DISTRICT OF COLUMBIA: Howard Ebenstein, District of Columbia Energy Office, 613 G Street N.W., Suite 500, Washington, D.C. 20001. 202-727-4700.

FLORIDA: Randy Zipser, Florida Governor's Energy Office, 214 S. Bronough Street, Tallahassee, FL 32399-0001. 904-488-6764.

GEORGIA: Rene Dziejowski, Office of Energy Resources, 270 Washington Street, S.W., Suite 615, Atlanta, GA 30334. 404-656-5176.

HAWAII: David Rezachek, Department of Business and Economic Development—Energy Division, 335 Merchant Street, Room 110, Honolulu, HI 96813. 808-548-4195.

IDAHO: Gerald Fleischman, Idaho Department of Water Resources, Statehouse, Boise, ID 83720. 208-327-7959.

ILLINOIS: David Loos, Illinois Department of Energy & Natural Resources, 325 West Adams, Room 300, Springfield, IL 62704. 217-785-3969. In-state toll-free: 800-252-8955.

INDIANA: Ed Robards, Office of Energy Policy, Indiana Department of Commerce, One North Capitol, Suite 700, Indianapolis, IN 46204-2288. 317-232-8969.

IOWA: Ed Woolsey, Energy Bureau, Iowa Department of Natural Resources, Wallace Building, Des Moines, IA 50319-0034. 515-281-7015.

KANSAS: Phil Dubach, State Corporation Commission, Research and Energy Analysis Division, State Office Building, Fourth Floor, Topeka, KS 66612. 913-296-5460.

KENTUCKY: John M. Stapleton, Kentucky Division of Energy, 691 Teton Trail, Frankfurt, KY 40601. 502-564-7192.

LOUISIANA: Joanna Gardner, Louisiana Department of Natural Resources, Energy Division, P.O. Box 44156, Baton Rouge, LA 70804-4156. 504-342-1298.

MAINE: Bruce Olson, Office of Energy Resources, State House Station 53, Augusta, ME 04333. 207-289-6026.

MARYLAND: W. Dale Baxter, Maryland Energy Office, 45 Calvert Street, Second Floor, Annapolis, MD 21401.

MASSACHUSETTS: David Diltz, Department of Energy Resources, 100 Cambridge Street, Fifteenth Floor, Boston, MA 02202. 617-727-4732.

MICHIGAN: Greg R. White, Michigan Public Service Commission, 6545 Mercantile Way, P.O. Box 30221, Lansing, MI 48909. 517-334-6422.

MINNESOTA: John Dunlop, Minnesota State Energy Division, 900 American Center Building, St. Paul, MN 55101. 612-296-4737. In-state toll-free: 800-652-9747.

MISSISSIPPI: Dianne Anderson, Energy and Transportation Division, 510 George Street, Jackson, MS 39202. 601-961-4733.

MISSOURI: Howard Hufford, Department of Natural Resources, Division of Energy, P.O. Box 176, Jefferson City, MO 65201. 314-751-4000.

MONTANA: Georgia Brensdal, Department of Natural Resources and Conservation, 1520 East Sixth Avenue, Helena, MT 59620. 406-444-6697.

NEBRASKA: Denise Disney, Nebraska State Energy Office, P.O. Box 95085, State Capitol Building, Ninth Floor, Lincoln, NE 68509. 402-471-2867.

NEVADA: Curtis Framel or David McNeil, Nevada Energy Office, Governor's Office of Community Services, Capitol Complex, Carson City, NV 89701. 702-687-4990.

NEW HAMPSHIRE: Scott Maltie or Barri-Lynn Medeiros, Governor's Energy Office, 2-1/2 Beacon Street, Concord, NH 03301-4498. 603-271-2711.

NEW JERSEY: Vincent Pedicini, New Jersey Board of Public Utilities Division of Energy Planning and Conservation, 101 Commerce Street, Newark, NJ 07102. 201-648-7005.

NEW MEXICO: Walter Smith, New Mexico Energy & Minerals Natural Resource Department, 2040 S. Pacheco Street, Santa Fe, NM 87505. 505-827-5907.

NEW YORK: Vicki Mastaitis, New York State Energy Office, Agency Building 2, Empire State Plaza, Albany, NY 12223. 518-473-0729.

NORTH CAROLINA: Larry Shirley, North Carolina Solar Center, c/o Energy Division, Department of Economic & Community Development, Box 7401, North Carolina State University, Raleigh, NC 27695-7401. 919-737-3480. In-state toll-free: 800-662-7131.

NORTH DAKOTA: Sherry Herman, Office of Intergovernmental Assistance, State Capitol, Fourteenth Floor, Bismarck, ND 58505. 701-224-2094.

OHIO: Abdur Rahim, Department of Development, 30 East Broad Street, No. 2400, Columbus, OH 43266. 614-466-6797.

OKLAHOMA: Steven E. Boggs, Oklahoma Department of Commerce, Division of Community Affairs and Development, P.O. Box 26980, Oklahoma City, OK 73126-0980. 405-841-9321.

OREGON: Gary Curtis, Oregon Department of Energy, 625 Marion Street, N.E., Salem, OR 97310. 503-378-8446.

PENNSYLVANIA: Phil Schuller, Pennsylvania Energy Office, 116 Pine Street, Harrisburg, PA 17101. 717-783-9981. In-state toll-free: 800-692-7312.

RHODE ISLAND: Roger L. Buck, Governor's Office of Housing, Energy and Intergovernmental Relations, 275 Westminster Street, Providence, RI 02903. 401-277-3370.

SOUTH CAROLINA: Jean-Paul Gouffray, Governor's Division of Energy, Agriculture and Natural Resources, 1205 Pendleton Street, Suite 333, Columbia, SC 29201. 803-734-0325.

SOUTH DAKOTA: Steve Wegman, Governor's Office of Energy Policy, 217-1/2 W. Missouri, Pierre, SD 57501. 605-773-3603.

TENNESSEE: Clinton A. Berry III, Tennessee Department of Economic and Community Development, Energy Division, 320 Sixth Avenue North, Sixth Floor, Nashville, TN 37243. 615-741-6671.

TEXAS: Judith Carroll, Governor's Energy Management Center, P.O. Box 12428, Austin, TX 78711. 512-463-1871.

UTAH: Britt Reed, Utah Energy Office, 3 Triad Center, Suite 450, Salt Lake City, UT 84180-1204. 801-538-5428.

VERMONT: Stuart Slote, Department of Public Service, Conservation and Renewable Energy Unit, State Office

Building, Montpelier, VT 05602. 802-828-2393. In-state toll-free: 800-642-3281.

VIRGINIA: Jennifer Snead, Department of Mines, Minerals & Energy, 2201 West Broad Street, Richmond, VA 23220. 804-367-6883.

WASHINGTON: Nancy Hanna, State Energy Office, FA-11, 809 Legion Way Southeast, Olympia, WA 98504-1211. 206-586-5021.

WEST VIRGINIA: G. W. Willis, West Virginia Fuel and Energy Office, 1204 Kanawha Blvd. East, Second Floor, Charleston, WV 25301. 304-348-8860.

WISCONSIN: Dan Moran. Wisconsin Energy Bureau, P.O. Box 7868, Madison, WI 53707. 608-266-1067.

WYOMING: Ed Maycumber, Economic Development Department, Energy Division, Herschler Building, Second Floor West, Cheyenne, WY 82002. 307-777-7284.

CHAPTER ELEVEN

Glossary

Terms and Key Words

Absorber plate: Surface in a flat-plate solar collector in which the solar energy is transferred to heat into the tubes running through it. The tubes are filled with either water or glycol, depending on the type of system.

Active solar system: Energy system designed to heat water or space and consisting of a solar collector, pump or fan, controller and water tank. Active systems are characterized by moving parts such as the pump.

Add-on system: Solar system which is retrofitted to an existing conventional heating system in a home, as opposed to being built into a new house.

Adobe: Clay found predominantly in the Southwest United States which is sun-dried and used for building materials. It has the properties of thwarting heat in summer and holding heat in winter.

Air-based solar heating system: Energy system where air is heated and distributed through the building to heat space, or less frequently, water.

Alternating current (AC): Standard household electricity in the United States where the flow of electrons reverses at regular intervals, usually 120 reversals per second or 60 cycles per second. The current flow from electrical outlets in your home is AC.

Annual load fraction: Percentage of the actual quantity of energy being replaced by solar energy.

Aperture: Opening or area of a solar collector or window through which solar radiation is admitted.

Azimuth (solar): One of the tools used to specify the sun's position at any given time. It is the angle due south at the horizon directly below the sun ("0" degrees) and angles east to west go to 180 degrees.

Baseload power plant: Electricity generating plant which supplies electric power on a continuous basis.

Berm: Mound of earth either abutting a wall to help stabilize building temperature inside, or positioned to deflect wind from the building.

Btu: British thermal unit, which is the standard unit for measuring heat energy. It is technically the amount of heat required to raise the temperature of one pound of water one degree Fahrenheit under stated conditions of pressure and temperature. One thousand Btus are approximately equal to the energy in an average candy bar, an hour of bicycling, two glasses of table wine, or four-fifths of a peanut butter and jelly sandwich. One million Btus of heat energy are equal to that of eight gallons of motor gasoline, or the amount it takes to move the average passenger car about 144 miles.

Building envelope: Materials of a building that enclose the interior spaces such as walls, roof, foundation, windows, etc.

Clerestory: Window located high in a house near the eaves which is used for light, heat gain, and ventilation.

Closed-loop solar system: Solar water heating system in which some type of liquid other than water (such as an anti-freeze solution) is heated by the sun. A heat exchanger then transfers this heat to the water.

Collector: Component of a solar system which collects sunlight and converts it to heat. It is usually an insulated box that is covered with some type of plastic or glass. It can also hold the water tank on some passive systems.

Concentrator: Curved reflector or lens which focuses sunlight onto either a pipe to collect heat or onto a solar cell to produce electricity.

Conductance: Measure or value of the speed in which heat is conducted through a material based upon a standard of which a Btu will flow through one square foot of material in one hour.

Conduction: The way heat moves directly through solid bodies. The most familiar of the heat transfer mechanisms, this is the method by which the handle of a metal spoon which is left in a pot of hot liquid is heated.

Convection: The way heat circulates through liquids and gases. When air comes into contact with an object which is warmer than it is, some of the heat is transferred to the fluid. As the fluid warms, it becomes lighter than the surrounding fluid and it rises; colder fluid is drawn in to replace it and the process continues until thermal equilibrium is reached. For example, warm air rises because it is lighter than cold air, which sinks.

Conversion efficiency: Amount of energy produced is expressed as a percentage of the amount of energy consumed.

Cooling capacity: Quantity of heat that a room air conditioner is capable of removing from a room in one hour's time.

Degree-day: Difference in degrees Fahrenheit below the average outdoor temperature for each day under the average indoor temperature of sixty-five degrees Fahrenheit. Therefore, the larger number of degree days indicates the severity of cold in the specific area. This is a good way to understand an area's climate and determine the type of cooling and heating systems needed.

Demand load: The necessary amount of energy for a particular need. For example, an air conditioner puts a certain demand load on the energy use in a house, as does the water heater or appliances.

Desiccant: Material which absorbs water and therefore dehumidifies. One popular use of desiccants is to package a small amount with new electronic equipment to keep moisture from damaging the equipment. A great deal of research is currently underway on the use of desiccants in both active and passive solar cooling systems.

Diffuse sunlight: Sunlight that reaches the earth after being reflected off of atmospheric particles and clouds.

Direct current (DC): This electrical current has a voltage that is unchanging over time and a one-directional flow of electrons through a conductor such as electricity from a battery. Photovoltaic systems produce DC current, which can be used or changed into AC by means of an inverter.

Direct sunlight: Light travelling a straight path from the sun.

Direct solar conversion: Process of changing sunlight to electricity via a solar electric cell (photovoltaics).

Direct solar gain: Passive solar design which allows sunlight to be converted to heat, usually through the south side of the building by the use of large windows and special building materials.

Domestic hot water system: Water heating energy system for a house.

Double glazing: Use of two layers of glass on windows, skylights, etc. This type of glazing traps air or gases that act as a better insulator than does thinner, single-paned glass.

Duct: Passageway made of sheet metal used for conveying cold or hot air. It sometimes includes a fan to propel air (known as a "duct fan").

Easement: Agreement to protect access to property. There have been a number of solar easement laws enacted in many states in recent years to protect homeowners' rights to solar access.

Efficiency: Ratio of energy output to the solar energy it intercepts which can be used for solar heating and solar electricity. The higher the efficiency of a system, the more solar energy it produces. If a solar cell is 10 percent efficient, it converts 10 percent of the light energy striking it into electricity.

Electrical energy: Flow of electrons in an electric current, usually measured in kilowatt (kW) hours. One kW equals 3,412.97 Btus.

Emissivity: A measure of how much heat a material gives off through radiation. Materials with low emissivity, such as radiant barriers, absorb very little solar radiation and reflect most of the heat back outside.

Energy: Forces which permit the capacity to do work using heat, electrical, kinetic, chemical, and radiant power.

Energy audit: Accounting for the forms of energy used (and lost) during a designated period. An energy audit of your home can help identify areas where energy efficiency can be improved or changed.

Evacuated tube collector: Collector composed of glass tubes which are filled with a vacuum. As a result they are usually more efficient, as well as more expensive, than conventional flat-plate solar thermal collectors.

Expansion tank: Tank which is part of a closed loop system which provides space for the expanded volume of liquid and gases which occur when they are heated.

f-chart: The name of a widely used solar system sizing procedure.

Glazing: Translucent material such as glass or plastic used for admitting light.

Header: A pipe that runs across the edge of solar thermal collectors which uniformly collects or distributes the heat transfer fluid and ensures equal flow rates and pressure.

Heat: Form of kinetic energy that flows from one place to another because of a temperature difference between them, usually measured in calories or Btus.

Heat exchanger: Device which exchanges heat from air-to-liquid, liquid-to-air, or liquid-to-liquid. The most common kind of heat exchanger used in solar is usually coiled pipe in a water tank.

Insolation: Amount of sunlight reaching a surface. It is technically the sum of diffuse and direct sunlight on a unit surface in a certain time, and is measured in Btu/square feet.

Integral collector storage system (ICS): Passive solar water heating system which combines the tank and collector in one unit. It is commonly known as a "breadbox" water heater.

Kilowatt (kW): Unit of power equal to 1000 watts. One horsepower uses .776 kilowatts. The typical U.S. home uses an average of 3 kilowatts of power for all of its electrical needs.

Kilowatt-hour (kWh): Unit of energy expressed in the use of kilowatts of energy for one hour. This is the standard way to express electrical energy usage.

Load: Amount of energy required or tapped at a specific time period.

Load profile: Distribution of energy needs over a period of time. This is a way of measuring how much energy is used in a home at certain time periods.

Lumen: Quantity of light to illuminate to an equal intensity of a one-foot candle. Lumens are the common standard of lighting output.

Mass: Density or weight of a material which is useful to determine how much heat can be stored in certain building materials.

Megawatt (MW): Electrical generating plant standard measure of one million watts or 1000 kilowatts (kW). The average nuclear power plant is 1,000 MWs.

Micron: One-thousandth of a millimeter.

Module: A framed panel with photovoltaic film or interconnected solar cells.

Movable insulation: Shades, blinds, and other window coverings that can be adjusted by the home's occupants.

One sun: Maximum natural sunlight that usually hits the earth at noon with no diffusion of sunlight.

Open-loop solar system: Solar system where the water is heated directly by the sun in the solar collector. No heat exchanger is needed since the water itself receives the sun's heat directly.

Parabolic trough collector: Reflector which is curved and focuses sunlight along a line. This type of collector is used in high temperature solar systems to produce very high temperatures for electricity and steam.

Passive solar design: Construction technique of using structural elements of a building to bring in heat during winter and deflect or vent heat during the summer through use of new materials, coatings, window sizing, eaves, building orientation, and landscape.

Passive solar system: Solar heating or cooling system that does not need the use of mechanical equipment such as fans or pumps to transfer the solar heat to storage.

Peak load: Time in which electric utility companies need extra power due to increased use of electricity for air conditioning, lighting, and other appliances. In many cases, this occurs during midday when solar is optimum.

Peak watt: Measurement unit for the performance of solar electric cells (photovoltaics) which will deliver one watt (W) of power under specified operating conditions with a solar insolation of 1000 watts per square meter.

Photoelectric effect: Absorption of a photon (light) by an atom which releases an orbital electron.

Photovoltaic (PV): Solid-state device which enhances the photoelectric effect whereby light is converted directly into DC current electricity.

Photovoltaic array: Interconnected set of photovoltaic modules which are integrated into one electric production system. Photovoltaic systems are modular, so arrays can be built as large as needed.

Photovoltaic cell: Semiconductor unit which converts sunlight to electricity. Many cells can be connected to a panel called a photovoltaic module.

Photovoltaic module: (See module)

Power conditioning: Changing or modifying the characteristics of electrical power, such as from DC to AC or from 12 volts to 120 volts using inverters or transformers.

Radiant barrier: Foil material used in airspaces in attics of homes, primarily in southern climates, to block solar radiation and keep a house cooler and more energy-efficient.

Radiation: Heat moving similar to light, in a vacuum or through a transparent medium. This is the only heat transfer mechanism which operates across a vacuum and is the primary means of energy transfer. Solar radiation is the way the sun heats the earth, and is the cause of environmental overheating and is most often the major source of building overheating. In a house, a warmed surface gives off heat radiation that travels toward a cooler surface.

Reflector: Shiny surface that reflects sunlight onto a receiver (pipe or oven).

Renewable energy resources: Energy that is virtually inexhaustible such as solar, wind, hydropower, geothermal, and biomass (growing and waste organic matter).

Retrofit: Subsequent installation or construction of a solar energy system after the initial building of the structure.

R-value: Measurement of the resistance of a material to heat flow, it is used to give the insulating quality of materials. The higher the R-value, the greater its insulating value so the better the material or insulation retains heat or cooling.

Shading: An effective way to keep a building cool using eaves, awnings, walls, trellises, and trees.

Solar cell: Photovoltaic cell.

Solar collector: Normally used to describe the device which traps heat from sunlight and transfers it to air or fluid. The term solar module is preferred for a solar electric (photovoltaic) collector.

Solar cooling: Use of solar heat in combination with heat pumps, gas or ammonia refrigerants, desiccants, or condensers. Solar cooling can also refer to the use of passive solar design to cool a building or photovoltaics to power air conditioners, refrigerators, or vent fans.

Solar electric cell: Photovoltaic cell.

Solar gain: Part of a building's heating load which is provided by solar radiation striking the building or passing into it through the windows.

Solar panel: Refers to either a photovoltaic module or a solar heat collector.

Solar thermal: Heat derived from sunlight which can be used directly for heating water or air, or through steam to generate electricity.

Solarium: Living space enclosed by glazing.

Thermal mass: Any material used to store the sun's heat or the night's coolness which is then released when needed and constitutes a part of a passive solar heating design.

Thermosyphon system: Passive system which has the solar collector and tank on the roof but needs no circulating pumps since the water, when heated, naturally rises into the tank.

Thin films: Refers to a thin layer of silicon deposited on glass to create a photovoltaic module. This is a type of solar electric material.

Time of day metering: When different electric rates are charged at different times of the day. It is based on the need of the utility company to charge more for energy during times of high energy consumption. Many utilities around the United States have adopted this type of system.

Transfer fluid: Fluid (usually glycol or water) used to absorb the heat from a solar collector in a closed-loop system.

Trombe wall: Wall used in passive solar design which absorbs heat and transfers it into the building.

Unglazed collector: Solar collector which has no glazing to trap the heat. This type of collector is typically used to provide heat for swimming pools.

Utility company: Either a privately-owned or city-owned company or a cooperative which produces electricity (also natural gas) and distributes it to the general public.

Vapor barrier: Material which prevents water from condensing on insulation and in walls.

Volt: Unit of electromotive force which indicates the intensity of electric current which is flowing.

Wafer: Thin piece of semiconductor silicon material used to make a solar cell.

Abbreviations

AC: Alternating current
Btu: British thermal unit
DC: Direct current
DHW: Domestic solar water heating system (for the home)
DOE: U.S. Department of Energy
FSEC: Florida Solar Energy Center
kW: Kilowatt
kWh: Kilowatt hour
MW: Megawatt
NATAS: National Appropriate Technology Assistance Service
PV: Photovoltaics
SEIA: Solar Energy Industries Association
SERI: Solar Energy Research Institute
SRCC: Solar Rating and Certification Corporation
Wp: Peak watt

Index